Rarely seen in captivity, Proboscis monkeys thrive at Twycross Zoo Park. Here are father, mother and infant

Bill Jordan and Stefan Ormrod

# The last great wild beast show

## A discussion on the failure of British animal collections

Constable London

First published in Great Britain 1978
by Constable and Company Ltd
10 Orange Street London WC2H 7EG


ISBN 0 09 461900 x

Set in Monotype Bembo
Printed in Great Britain by
Ebenezer Baylis and Son Ltd
The Trinity Press, Worcester, and London

To our children,
who will inherit our decisions

# Contents

# List of illustrations

# Introduction

Why have zoos? How can the incarceration of wild creatures for the mere amusement of human spectators be justified? Is the whole concept an abomination that has survived from an earlier stage in our history? Would society do well to remember Blake's line 'A Robin Redbreast in a Cage, puts all Heaven in a Rage', and outlaw zoos altogether?

We have both spent the majority of our working lives involved with wild animals and their welfare, in particular with those kept in captivity, so it is not surprising that we have become natural targets for this hysterical type of cross-examination. To establish a clear understanding of a zoo's role in the modern world, we should, at the onset of any such discussion, make a clear distinction between the zoo and the circus, for, surprisingly enough, many people still consider these two quite separate organizations as synonymous. The idea is of course absurd and one would be hard-pressed to find any genuine animal lover who would give support to the circus on any grounds. It can be objected to for many reasons: cruelty does undeniably exist in training, the travelling animal wagons are always foully cramped, and ultimately it has no purpose other than to amuse an apparently witless audience.

But a zoo is an entirely different concept. First, it cannot be viewed simply as a place where wild animals are locked up for the amusement of humans: that image belongs to the Victorian era. The zoo has a far more serious role to play, and its primary functions are concerned with research, education and vital breeding programmes associated with the preservation of those species gravely threatened with extinction in the wild. It might be thought of as a sort of Noah's Ark, complete with laboratories and classrooms.

As for the issue of confinement, this is largely a matter of degree.

A robin redbreast in a cage might understandably put all Heaven in a rage, but before a zoo biologist passed judgement he would want to know the size, shape and quality of that cage. A conventional goldfish bowl or parrot cage, for instance, should make Heaven rage because they can be classified as precise cruelty (even though legal), whereas some of the better aquariums and free-flight aviaries to be seen in zoos might be expected to placate even the severest critics.

No wild animal is free in the sense understood by the average citizen. Each living creature is confined by a maze of restrictions. These may be environmental, or may involve the presence of other animals or members of its own species, or the availability of food. If these factors are taken into consideration and the animal's normal biological needs are catered for, what does it suffer by being captive? As our scientific knowledge improves, so does the lot of the zoo animal. It is infinitely better off than its wild counterpart (and also the average pet hamster), for it has a regular, well-balanced diet, constant shelter, no territorial disputes, plus the benefits of the most up-to-date advances in surgery and medicine. Is it to be wondered at that the zoo specimen can expect a longer life than it would in the wild?

We have presented this defence of the institution many times and have occasionally had the rewarding experience of seeing a bitter opponent of zoos, realizing that their function is one of promoting a respect for life rather than denigrating it, become converted.

But in recent years doubts have begun to cloud our minds and, even as the well-prepared phrases have passed our lips, the hollowness of the argument has been there. Gradually we have had to come to terms with the fact that we are merely presenting a credible hypothesis of the functions of the modern zoo, and that it bears little resemblance to the present state of affairs in Britain.

To start with those three justifications for zoos: research, education and conservation. Hardly a zoo in the British Isles actively undertakes or sponsors research of any kind. Of those that provide educational programmes, the majority that seem to do so are really only trying to appear respectable and, thus, hardly merit serious consideration. As for conservation: on a world-wide basis only five species can be considered saved from extinction by captive breeding, and even one of these was a chance success. Nevertheless, there are notable exceptions. Several British zoos devote an incalculable amount of time and money attempting to conserve just one species and they have already made significant contributions. It is only when these contributions are compared against the total number of zoos in the country that they pale to insignificance.

Quite apart from these factors, one only has to visit a few of our wildlife collections to see that something is clearly wrong. Our own numerous visits to zoos have produced little that can be put forward in their favour and we estimate that about 90 per cent of those we have inspected are keeping some, if not all, of their animals in a state of deprivation that must result in physical, emotional or mental illness. Considered in this light, the zoo immediately assumes the role of an animated fun-fair. But it is too easy to produce an argument for the abolition of zoos, bearing in mind the neglect and suffering so often seen in them, and although it might seem expedient simply to arouse hostility against an apparently loathsome institution, it is more constructive to show that that same institution has potential and can be of unique value to man and wildlife. Indeed, the survival of the zoo has become essential if only because for some species captive breeding is the only alternative to extinction.

The root cause of the present lamentable state of our zoos can be traced to the basic rules of economics. Zoos are costly to run, so costly that if the needs of the inmates are to be adequately catered for, the zoo does not present a viable business prospect. Yet all too often the main reason for opening a new zoo is to make money, and this can easily be achieved if the zoo does not intend to carry out scientific research, organize educational programmes or specialize in breeding endangered species. With an estimated 20 million visitors a year clamouring to get into the gates, one can hardly express surprise at the commercial exploitation of the zoo, or that so many have opted for a share in this newly-discovered business. For the zoo boom is a fairly recent development.

At the end of the Second World War there were only 14 zoos in Britain. The number slowly increased throughout the 'fifties until in 1959 the total had reached 31. In the next decade almost 100 animal collections were added to this figure. This means that on average another new zoo opened every six weeks. Clear evidence of the success of a purely commercial approach to running wildlife collections can be seen in the phenomenal rise in the number of safari parks. The first of these was Longleat, which opened under the partnership of Jimmy Chipperfield and Lord Bath in 1966, and Chipperfield has since followed it with more, including Woburn, Knowsley, Blair Drummond, Lambton, the West Midland Safari Park and others in several parts of Europe. Another circus group, Smart Brothers, owners of Billy Smart's Circus, opened the Royal Windsor Safari Park in 1969 (sold to Trident Television, May 1977) and that year also heralded the arrival of a new business group on the zoo scene: Scotia Investments Limited. Mainly concerned with bingo halls and package holidays, this group managed to obtain eight zoos in its first year of operation. Chessington Zoo was purchased by the Spearshaft Industrial Group and Fortes Holdings Limited acquired Belle Vue in Manchester.* Piece by piece zoos were becoming absorbed into the leisure industry.

As the incentive behind nearly all zoo development in Britain in the last 15 years has been commercial, there has been an obvious shifting of values. Zoological objectives and animal welfare are of less relevance on a balance sheet, so the attitude towards the stock has become dispassionate, almost callous. For many zoo operators, the wild animals are simply goods placed on display in return for hard cash.

The puzzle, therefore, is why such an intolerable state of affairs

* In August 1977, Fortes announced that the zoo was to be closed owing to the enormous financial losses it had experienced. Belle Vue is now to be solely an entertainment centre.

has been allowed to develop. The laws which relate to the keeping of wild animals in captivity are inadequate. Anyone, providing they can obtain a plot of land and a few animals, can open a zoo. All that is required is planning permission from the local authority. But why such a situation has been allowed to continue is more difficult to understand.

The old-established zoos reacted to the sudden mushrooming of competition by creating the Federation of Zoological Gardens, an organization intended to establish minimum standards of care, housing, feeding, hygiene and safety, and sound principles of animal management. Membership is optional, but those wishing to join are obliged to conform to certain standards before being granted admission. Several commercial groups objected to this and claimed that all zoos should be incorporated into the Federation before pressure was applied to improve conditions; one group declined to join because they felt that the Federation was 'an organization biased in favour of scientific research'.

In 1971 Lord Janner introduced a Bill in Parliament for the Control of Zoological Gardens and at about the same time a rival organization to the Federation, namely The National Zoological Association, was established. It was all that was needed to make the natural antipathy between the scientific and commercial zoos erupt into a fervour of hostility. The Association's membership was soon twice that of its rival, presumably giving some credence to the supposition that 'might is right', for the Government finally abandoned the Bill, suggesting that the zoos of Britain settle their differences of opinion before returning with new proposals that would be acceptable to everyone. To this end a Central Zoo Union has been set up to provide an area of neutral ground between the two groups. All this really seems to mean is that agreements might be reached without causing anyone to lose face publicly.

In our opinion Lord Janner's original Bill could have included stricter controls, but the next set of proposals will, whichever way one looks at them, most probably be watered down to suit all palates.

Meanwhile, in the midst of all the petty arguments, discussions and backbiting, many animals remain neglected and live in conditions varying from poor to what can only be described as absolutely inhumane. As captive animals are obviously not captive by choice, we feel they are entitled to have their needs met as fully as the present state of knowledge allows, and that the only way to realize this ideal is to develop, as quickly as possible, the attributes of the good collections and close down the multitude of animal concentration camps dotted across Britain. If this does not happen, the day could easily come when public opinion turns against the zoo altogether; as an institution it could become as endangered as the species it purports to protect.

It was with such thoughts in mind that we decided to write this book: to attempt to communicate with some of those 20 million visitors and persuade them to be a little more discriminating in their patronage; to explain how zoos started, how they should develop and how damnable they can be when they lack any objectives save making money. Finally, we hope that some of those visitors might be prompted to act on behalf of those creatures that are clearly victims of the less savoury collections. The layman is not helpless in these situations: providing he knows what to look for and what questions to ask the keepers, he can quickly gain a fairly accurate picture of the zoo's validity. If not satisfied, he can complain to the staff and/or the management of the zoo. He can also write to a number of animal welfare organizations, the local council and the local Member of Parliament, preferably all three. Public opinion is a powerful force, so the more people who react when they see cruelty or neglect, the quicker it can be halted. Of course, some will say that these problems can only be resolved by the zoo organizations of this country. To these people we would say that an element of pressure would not be unreasonable. After all, the British zoo has taken 150 years to get to its present state (the first 'zoo' opened in 1826), and in far too many cases, this is virtually nowhere!

Chapter 1

# The bounty of which age?

To see a strange outlandish Fowl,
A quaint Baboon, an Ape, an Owl,
A dancing Bear, a Giant bone,
A foolish Engine move alone,
A Morris-dance, a puppet play,
Mad Tom to sing a Roundelay,
A woman dancing on a Rope,
Bull-baiting also at the Hope;
A Rhymer's Jest, a Juggler's Cheats,
A tumbler showing cunning feats
Or players acting on the stage,
There goes the bounty of our Age.

*Traditional*

When the plan of this book was first agreed it was decided not to include a history of zoos for two reasons: not only does every zoo book start with a potted history, but also as this one was intended to deal solely with present-day problems and not those of the past, a history seemed superfluous. Several colleagues criticized this decision, the intimation being that the general reader could not possibly comprehend how easily the zoo can veer from being an institution of intrinsic value to simply a cheap sideshow, without first being acquainted with its ancestry. Nevertheless, we resisted their advice and concentrated on the factors that seemed relevant—for space in a book is extremely valuable.

However, in our efforts to keep the text up-to-date we stumbled across an animal collection which subsequently made us change our minds. In the spring of 1976 we visited several establishments in a sort of package tour of zoos, but one particular zoo was of special interest as it had aroused a considerable amount of hostile criticism and we were naturally curious as to the truth behind the allegations. If anything, our informants had understated the atrocious conditions in which the animals were forced to live. Cages were cramped and filthy and a number of the inhabitants behaved in a manner that hinted strongly at mental maladjustment.

It was not the sights, sounds and smells alone, however, that caused the change of heart, but the guide-book that accompanied the collection, which contained the following obsequious declaration:

> Today zoos are not peep-shows; they are scientific establishments where the behaviour of animals is studied. The knowledge so gained helps to save wild animals facing extinction.

This so-called zoo, like many others, had little compunction about [intentionally] misleading its visitors as to the purpose behind its operation, and moreover it had the gall to imply that it was a scientifically run collection which bore little or no resemblance to the hideous menageries of the past. Yet in fact the opposite was true.

That hypocritical little preface clearly showed that a certain amount of historical background was important, if only to demonstrate how many of the less desirable human attitudes to wildlife have survived the centuries and how, in essence, the Victorian menagerie is still alive and well in Britain today. For all the very best things to be found in the modern zoo emanate from a philosophy concerning the rights of living creatures, and the need for a greater respect and appreciation of life by mankind as a whole. Such altruism is rare enough even in this society, but in the story of the zoo's ancestry it hardly features at all.

### *The first menageries*

The term 'zoo' was not coined until 1826 when the foundation of the Zoological Society of London provided the English-speaking world with a new word for an ancient concept. Menagerie is generally used to describe animal collections preceding the eighteenth century, and since these were usually crude, barbaric places, the term is also applied in the derogatory sense to present-day collections if they are old-fashioned or sub-standard in any way.

Throughout the ages, man has consorted with wild animals in no small degree and his practice of keeping them in captivity can be traced back for more than 3,000 years. The private collection of King Shulgai on his royal estate at Puzrish, Mesopotamia, is probably the earliest authenticated menagerie, although, doubtless, even earlier rulers had hunting parks of some kind. Hunting played a major part in the development of zoological interest; as civilization spread, wild beasts ceased to represent direct competition to mankind and little by little acquired a curiosity value. Hunting accordingly changed: instead of being a straightforward quest for food, it turned into a pastime. Killing for fun was provided with elaborate rules of fair play (at least from the human's point of view) and, under the blessing of royal patronage, gradually acquired the identity of a noble sport. Herein lay the foundations of the game reserve and, indirectly, the zoo. Royal parks were set up and the game animals indigenous to the area were carefully managed to ensure that a plentiful supply was available at all times. If the king and his guests wished to hunt—the game would be provided at all costs.

As new lands were discovered and conquered, so were new species of animals found, and it became a common practice for explorers and military leaders to return from their travels laden with exotic creatures which would hopefully please their king. Understandably few of these animals could be released into the protected areas, as they were either too ferocious, too timid, or unsuited to the environment, so cages were constructed to hold them securely. The menagerie was begun.

King Shulgai established his menagerie of exotics, as did many of the Egyptian Pharaohs, and in all probability the vast majority of these animals were accumulated as a result of gifts from foreign rulers or their emissaries. Such gifts are not uncommon even today. Most zoos contain animals that have been donated by admirers and 'friends of the zoo' and the custom of presenting wild animals from one national leader to another as a pledge of friendship is as frequently seen in this century as it was in the earliest history of civilization. Nehru gave elephants to visiting diplomats on an embarrassingly large scale, Khruschev sent Princess Anne a bear cub, and more recently the People's Republic of China presented The Right Honourable Edward Heath with a pair of giant pandas, which are now exhibited in London Zoo.

It does not follow that these owners of wild animal collections were necessarily fond of animals. The possession of an exotic creature was considered a sign of prestige, power, a status symbol comparable to the present-day kudos of owning a Rolls Royce, because only the wealthiest in the kingdom could afford to maintain these curiosities.

The pastime seems to have developed into a business at an early stage. Correspondence between King Burnabish of Babylon and the Pharaohs of Egypt regarding the purchase of animals for his royal menagerie still survives. Animal collecting at this time was not always a casual affair, not merely a chance discovery and capture by one of civilization's travellers. We know, for instance, of a truly mammoth wildlife safari that was organized by Queen Hatshepsut in 1490 BC. This expedition consisting of five ships ventured as far afield as Somaliland and returned laden with hundreds of birds, various monkeys, wild and domestic cattle, leopards and a giraffe, all of which were housed in the Queen's palace menagerie.

Giraffes were later to become quite fashionable with Egyptian royalty. Rameses II had a number of them in his large menagerie and he also had what is probably the earliest recorded case of a pet lion. The lion, who was called Antam-nekht, accompanied his master everywhere, including on to the battlefield, and lived chained in front of the Pharaoh's door for the remainder of the time.

At about the same period wild animal collections were developing in India, and in China the Emperor Wen Wang established one of some 1,500 acres, which he named the Ling-Yu, or Garden of Intelligence. Wen Wang's Garden can be dated from about 1000 BC and must have remained in existence for a great length of time. It was certainly copied by his descendants and by the time Marco Polo (1256–1323) visited the Great Kublai Khan there were many such Gardens of Intelligence spread throughout the land, all the personal property of the Emperor. Marco Polo was more fortunate than the natives of Kambaluk (Peking) in that he was allowed to see the fascinating specimens in the main collection; ordinary citizens were not admitted.

The Ancient Greeks were keen pet owners, and wild birds, monkeys and apes were quite common household pets as early as the fourth century BC. But the Greeks provide a pleasant interlude, an age of enlightenment, in the history of the zoo, for they took wildlife more seriously than any civilization had before, or would again for several centuries. Education, so often heralded by our modern zoos as one of their major contributions to society, was a key function of most of the city menageries and there is good reason to believe that the students in Athens were taken by their tutors on zoo tours 300 years before the birth of Christ.

Zoology became a serious science. Aristotle was one of the first men to make use of captive animals for scientific study and his *History of Animals*—the world's first encyclopaedia of zoology—was probably largely compiled from observations made in the menagerie of Alexander the Great. Alexander had been a pupil of Aristotle's and had obviously gained much of his own enthusiasm for the natural sciences from his beloved old tutor. So in an age when the potentates of Asia Minor were beginning to stage arena animal fights, the Greeks displayed a real change in the motives of zoo culture by developing a more studious approach towards animals. Yet as the sun set upon the Greek Empire, the zoo's prospects dimmed also.

### *Bread and circuses*

The emergence of Rome as a great civilization forms a baffling episode in zoo history. Throughout the whole period of the Empire's existence, many of the Roman aristocrats established private menageries at their villas and presumably the high cost of maintaining these engendered concern as to the well-being of the animals. There certainly existed a humane attitude towards animals, at least amongst zoologically-minded intellectuals of the time, many of whom had some very forward-looking ideas about wild animal husbandry. For example, it seems likely that the walk-through aviary was originated by one of the more inspired Roman collectors.

Pliny the Elder (AD 23–79) was the greatest of Rome's natural historians and his book called simply *Natural History* was the first work of any stature since that of Aristotle. Even by modern-day standards it is an outstanding document. But the zoological zeal that so obviously flourished in the heyday of the Empire makes it extremely difficult to understand how that same society could nurture such an insatiable lust for bloody spectacle, an enthusiasm that culminated in the Roman circus. Yet the two levels of interest in wild animals did exist, on the one hand there were the private bird collections and menageries, whilst on the other there were the amphitheatres.

The infamous Roman Games were no short-lived affair. They lasted for almost 500 years and undoubtedly contributed to the extinction of a considerable number of species. Although we will never know how many hundreds of thousands of wild creatures died upon the sands of the Colosseum, Circus Maximus and innumerable smaller arenas dotted across the Empire, the carnage was unquestionably massive, and by the time of Nero's reign several species, like leopards, were already in short supply. This shortage was reflected in the cost of putting on the games: as the collecting expeditions had to go further and further afield in order to obtain sufficient numbers of beasts for the arena, so the cost rose. Nevertheless, Nero's successor, the Emperor Titus, (to whom Pliny had dedicated his *Natural History*) still managed to put on a public spectacle that involved the deaths of 5,000 animals in one day and a further 9,000 in the following three months.

Beast shows were originally introduced as a subsidiary attraction to the gladiatorial combat, but in time this supporting feature seems to have become more popular than the chariot-racing, swordsmanship and athletic events which had formerly constituted the main programme. These spectacles included events like lions fighting bears, elephants against rhinoceroses, and, of course, men (both armed and unarmed) versus a variety of beasts. All this was accepted by the authorities on the grounds of giving the public what it wanted. But continual mass slaughter eventually palled and new ideas had to be thought up to keep the populace amused, so the animals were trained to perform tricks in an effort to vary entertainment; bears were taught to dance, lions jumped through hoops and elephants attempted ludicrous balancing acts. Even after the establishment of the Holy Roman Empire the games continued. The inclusion of human combatants was stopped and men were only allowed to compete in relatively non-injurious competitions, but wild animals were still pitted against each other in futile trials of strength and aggression or made to perform tricks. Not surprisingly a society so used to seeing the massacre of innocent human victims expressed little concern for what it considered lesser creatures, and so what had begun as a supplementary feature of the amphitheatre turned into a wholesale extermination programme executed under licence of the mob.

By the end of the games large tracts of North Africa, the Mediterranean and Atlantic coasts had been irreparably depleted of native fauna, so it seems likely that the private menageries of the Roman naturalists were finally plundered of their stocks in a last desperate effort to fuel the arena and provide an already restless people with diversion in the form of an extravagant and bloody drama. The attempt failed as did the Empire.

Although we have only touched upon the historical facts of the period, we cannot gloss over the significance of the Roman contribution to social attitudes about wildlife. Rome's intellectuals outlined important principles for the management of wild animals in captivity; in fact a fantastic amount of new ground was broken during this time and the Romans laid some vital foundations for the development of the zoological park. Yet, whilst society's élite furthered the Greeks' vision of the study of the natural world, the amphitheatre promoted a distinctly opposite view of wildlife—an attitude that has persisted right up to the present century and is currently evidenced by the continued existence of the circus and the bullfight. This influence was clearly displayed when Theodosius II opened a public animal park in AD 425 in Constantinople, for he incorporated an arena in it.

### *The Middle Ages*

Charlemagne contributed significantly to the development of wildlife parks by establishing several on monastery lands, the monks being made responsible for their day-to-day organization and the welfare of the animals. Considering the Church's opinion of animals at that time—animals were deemed to be without souls and could easily be possessed by evil spirits—it is perhaps unfortunate that no record survives as to the reaction of the various abbots upon receipt of the imperial orders that were to make so many of their monks full-time animal keepers. Whatever it was, it seems that they were highly efficient at their appointed task, for these monastic menageries were used as models for many of the European royal collections centuries later.

By the Middle Ages the craze for owning wild animals had grown. Civilized man was now travelling further across the globe in a concentrated drive to subdue new continents and open fresh areas of trade. Understandably in the process he was also discovering strange, exotic creatures unlike anything seen before. Central Europe and Asia became the spawning grounds for menageries, areas of high population with large estates owned by powerful, rich and educated men with varying interest in wildlife.

In England the royal menagerie was originally established by Henry I at the mansion built by his father, William the Conqueror, at Woodstock in Oxfordshire. Henry III moved the animals to the Tower of London, where the menagerie remained for over 500 years without ever becoming a collection of any great merit. The royal collection was in fact the constant victim of the whims of successive monarchs, its fortunes fluctuating from one reign to the next. In Henry III's time its future must have looked fairly promising, as many new animals were donated by the Holy Roman Emperor, Frederick II, who himself owned a large travelling menagerie and three private collections in Italy. But in the following years the menagerie languished, was revived, then allowed to languish again until by the Tudor era it had become a revolting parody of the Roman Games, with court guests being entertained with staged animal fights—blinded bears were pitted against dogs, lions against tigers, and, of course, there were the inevitable cock-fights.

A menagerie of sorts was maintained at the Tower right up until the formation of the Zoological Society of London when the few remaining animals were removed to Regent's Park. For a time an entrance fee was charged, but by the mid-eighteenth century it had become a common practice for the poorer Londoners to take their unwanted pets in lieu of money. These pathetic creatures (generally cats and dogs) were gladly accepted by the gateman as they made a saving on meat purchases. Also, as they were usually pushed into the carnivores whilst still alive, he probably obtained some substantial tips from the more affluent members of the audience.

But to return to the Middle Ages. The story continues along a similar vein throughout the developing civilizations and the list of aristocrats and royalty with a taste for animal collecting grows longer and longer with each decade. Sultans, kings, emperors and popes indulged their passion for menageries, and even notable men like Leonardo da Vinci seem to have kept small collections, which were used as models for sketches and paintings.

Remarkably, although the institution has suffered badly from inherited influences, there is irrefutable evidence that the 'zoo' concept developed quite independently within cultures that could not possibly have copied the ideas of another. The most positive proof of this was the astonishing discovery made by Cortés in 1519 when he and his *conquistadores* arrived at the city of Tenochtitlán (now Mexico City) and were shown a menagerie of a size and splendour unknown in the western world. This menagerie, owned by Montezuma, had a staff of over 400 keepers and was the largest wild animal collection ever accumulated—even by present-day standards, it was vast. There were ten large ponds upon which lived a myriad of waterfowl; hundreds of reptiles were kept in sunken vivariums; the carnivorous animals were housed in cages constructed of bronze bars; and aquariums containing the Emperor's huge collection of fishes stretched as far as the eye could see. It is unfortunate that detailed illustrations of the design features have not survived as it seems likely that these might have shown a zoological park that was literally centuries ahead of itself. The Spaniards record that the menagerie was staffed by a team of nurses whose sole duty was to care for the sick animals (this is more than can be said of many zoological parks today); dozens of workmen prepared the daily foodstuffs; others were merely employed to collect the fallen feathers from the aviaries for use on the many ceremonial robes worn at court.

Cortés was disturbed to find human freaks amongst the collection's 'zoological specimens', but from a safe vantage point of historical distance we find that similar barbarity was to be found in Europe. Italy had an exclusively human menagerie filled with slaves, and many of the earlier collections exhibited various races of *homo sapiens* alongside other species; the Roman Amphitheatre is the obvious example that readily springs to mind. In the nineteenth century Münster Zoo proudly displayed three iron cages containing the effigies of men who had been tortured to death, and more recently a safari park organization considered establishing an African Village, incorporating gift-shops, a restaurant and negroes in traditional costume!

Unfortunately, two years after finding Montezuma's incredible animal park, the Spaniards attacked the city and during the ensuing siege the defenders were forced to kill and eat the animals of the royal collection. Finally, the conquerors razed the city to the ground, including the palace and the royal menagerie.

The conquest of South America soon had its impact on the animal collectors. Ferdinand of Aragon was the first to acquire a collection of South American fauna at his palace at Toledo, but it was King Manuel I of Portugal who controlled the market in exotic animals during the following years, keeping South American monkeys and macaws, as well as elephants, cheetahs, lions and baboons in his royal park. He appears to have been something of an animal dealer as well as a collector, selling animals to wealthy merchants throughout Europe and constantly entertaining them with displays of his wild beasts in an effort to justify the high and continually escalating prices he was asking for them. It was he, incidentally, who imported the first live rhinoceros into Europe since the Roman Games, a creature whose main claim to fame was to be its immortalization in the woodcut executed by the great Nüremberg painter Albrecht Dürer which for many years was considered the only authentic picture of the species. Today this portrait's real beauty lies in its naïve interpretation of a species we now know as the great Indian rhinoceros (*Rhinoceros unicornis*), rather than its pretence at zoological accuracy.

Rather typically, Manuel chose to pit this rare giant against an elephant in the arena, but as neither animal showed any inclination to fight, he decided to present his rhinoceros to the Pope; the ship transporting the animal sank before reaching its destination.

### *Akbar the Mogul*

In the sixteenth century we find one ruler who cannot be casually dismissed as yet another menagerie fanatic, a man who has been described by historians not only as a remarkable zoologist, but as the greatest ethical teacher since Marcus Aurelius. Jalal ud-din Mohammed Akbar was the greatest and wisest of the Mogul emperors. His economic and social reforms resulted in the unification of a vast empire, and his wildlife parks set an example to the rest of the civilized world.

By the end of the sixteenth century Akbar had established wildlife collections in many Indian cities, but these were not like the cramped royal menageries that were developing in Europe: they were large areas containing spacious enclosures and cages. Each one had resident veterinary officers so as to ensure that medical attention was available for the animals at all times. Breeding and the study of animal behaviour was enthusiastically encouraged. Arena fights, baiting or cruelty of any kind were strictly forbidden because Akbar considered it beneath man's dignity to inflict suffering on his fellow creatures. He devoted himself to the task of instilling into his people a love and respect for wildlife. When the gates of his many nature reserves were opened to the people of the empire they were found to contain thousands of marvellous animals—leopards, tigers, monkeys, cheetahs, rhinoceroses, various deer and small mammals, exotic birds and over 5,000 elephants. And Akbar's message to the visitors to his collections was simply: meet your brothers, take them to your hearts and respect them. This Mongol tradition of love for animals has remained an important aspect of northern Indian Muslim culture to this day, largely due to the influence of this one man. Akbar was a new breed of naturalist, a genuine zoophilist who lacked the pretentiousness of his counterparts in other lands, for he owned these animals for the joy of all his people as much as for himself, not because they could be used as symbols of power and riches.

### *Developments in Europe*

A century after the opening of the Indian Imperial collections, King Louis XIV of France also decided to display his private collection of animals to visitors and in 1663 he converted the Jardin du Roi in Paris into a zoological garden—although officially common folk were not admitted, in practice it was not possible to keep them out. It is unlikely that Louis was particularly interested in animals, but he was born into an age when menageries were fashionable in court circles. His passion was architecture, so he combined the two subjects to provide himself with a hobby which resulted in a royal menagerie that was to inspire visitors from many nations to embark on similar ventures. Louis was also stubborn, self-centred and fickle: soon losing interest in his zoological developments, he allowed them gradually to fall into disrepair and decay.

However, Louis XIV had inspired the use of new techniques in exhibiting wild animals, and many of his innovations, such as the use of trees and flowering shrubs to screen off enclosures, are as valid today as when he first used them, but the menagerie at Versailles was, in reality, a more ostentatious than practical zoological garden.

At the end of the eighteenth century, as the French aristocrats fell before the guillotine in the revolution, so their follies fell to the mob, and many of the menageries were pillaged and animals slaughtered, thus bringing to an end several of the larger European private collections.

Some of the animals that survived the revolution were finally taken to the Royal Garden of Medicinal Plants, the Jardin des Plantes, where the famous naturalist Cuvier was appointed to re-organize the new collection and the Committee of Public Safety ordained that the menagerie was henceforth to be open to the people, admission free. Apart from small charges for entrance to some of the houses, it still is. Famous scientists of the age all made use of the Jardin, Buffon had worked there prior to the arrival of the animals and Cuvier, Lamarck, and Latreille all used the living specimens to further their researches into natural history. Among the interesting species kept in the new menagerie were an Indian rhinoceros (*Rhinoceros unicornis*) and a quagga (*Equus quagga*), both having been rescued from the valuable collection of Versailles. Later the Emperor Napoleon added a considerable number of species to what was rapidly becoming the leading academic zoological garden in Europe. In 1841 the Jardin des Plantes had a library of 28,000 books, botanical and geological museums, a zoological museum and a museum of comparative anatomy, the majority of which was destroyed in the Siege of Paris in 1870. Today the Jardin is a disappointing collection.

At the beginning of the eighteenth century the natural sciences began to take on a more definite shape. Man's attitude towards animals had rarely been altruistic, but the age of the great explorer-naturalists was starting to bear influence on the academics of the day. Carl Linnaeus, perhaps the world's most famous naturalist (1707–1778), took upon himself the incredible task of classifying and naming the innumerable species of animals and plants in the living world. So great was his success that the Linnear system of using two Latin words to describe each species is the basis for all modern zoological nomenclature. It is believed that the later-formed Linnean Society was partially responsible for inspiring the inauguration of the Zoological Society of London. Armed with this remarkable system of classifying their finds, the great explorers and scientists rushed into a fever of natural history researches. Buffon, at one time custodian of the Jardin du Roi, embarked on his monumental *Histoire Naturelle* which was finally issued in 45 volumes, and became so fashionable that it was to be found in most of Europe's elegant drawing-rooms. John Gould's 25 volumes of bird illustrations found equal success. Sir Joseph Banks sponsored a young naturalist, Robert Brown, on a voyage to the Australasian regions, and Baron Alexander von Humboldt boldly explored the riches and not inconsiderable dangers of South America's secrets. It was a flourishing age for the natural historian and it was therefore to be expected that decisive changes would appear in the European zoological and botanical collections —not just unusual specimens, but new academic approaches.

When London opened its gates to the public in 1828 (two years after the formation of the Zoological Society) an entirely new concept was thrust upon the British public. The word 'zoo' was first honoured by appearing in the *Oxford English Dictionary* in 1847, although it probably seemed more significant to the general public when it was featured in a popular music-hall song called 'Walking in the Zoo is the O.K. Thing To Do'.

### *The nineteenth century*

The Zoological Society was the first gathering of learned men whose prime objective was the serious study of wild animals. This makes it, historically, the most important of the world's zoological parks. Four years later the Dublin Zoological Society was founded, and both London and Dublin inherited the remaining animals from the Tower menagerie, which was finally dispersed in the 1830s. The Tower menagerie had by now declined to conditions of abysmal squalor, even though George II had enlarged the collection and made its curiosities available to the citizens of London. Bristol opened its gates at about the same time, followed by Amsterdam in 1837, Antwerp in 1843 and in 1844 King Frederick Wilhelm IV of Prussia donated his private collection to the people, under the directorship of the most celebrated naturalist of the age, Alexander von Humboldt. Berlin began its first major expansion scheme in 1869 under the direction of the distinguished zoologist, Dr Heinrich Bodinus, and by 1890 they had built aviaries, lion houses, a camel house, ungulate enclosures, deer-houses, a monkey-house and three restaurants for the visitors, who by the turn of the century often amounted to as many as 60,000 on a fine day. Even more important, the off-view amenities included site drainage, pump-houses and refrigerated units for the animal feedstuffs. With such a history, it is not surprising to find that the modern West Berlin zoo is one of the finest in the world.

The list continues: a new zoo in Hamburg in 1860, in Budapest in 1865 and in Basle in 1873. The Edinburgh zoo had opened in 1839, but collapsed 20 years later, and it was not to be resurrected until 1931.

Although these new Society zoos were, superficially at least, institutions for the advancement of the natural sciences, the overall policy seems to have been to cram as many different species of animals into barred cages or pits as was possible, so that in spite of their numerous pretentious charters of zoological intent, they were basically little better than their predecessors of the Middle Ages. There were two main reasons for this. First, zoo biology as a science did not exist. No one stopped to consider the biological requirements of a specimen when housing it; the zoo was looked upon as a new type of natural history museum which should exhibit as many specimens as possible. And, although security was recognized as a feature of great importance, few original restraint techniques appeared in the new zoological gardens, most enclosures reflecting the methods employed in the menageries of the ancient world. Potentially dangerous animals were shut securely behind strong metal bars in cramped and, consequently, insanitary conditions, and tropical species which were considered less hardy than the more commonly kept species were virtually suffocated in hot, badly ventilated animal houses.

The profusion of bear pits to be found in zoos at this period (bears have always been the commonest animals in zoos and circuses) were all modelled on the famous bear pit that was established in Berne, Switzerland, in 1513 and even then it had not been a new idea. Animal pits had originally been developed as pit-traps by the early Neanderthal hunters! The bear pit at Berne still contains bears—the bear is the city's emblem—and is thus the oldest site still in use for wild animals in Europe.

Zoo architecture flourished, but the attention to detail was reflected in the impressive façades of the zoo entrances and animal houses rather than the suitability of the animals' accommodation. The designers took much of their inspiration from the cultures of the ancient civilizations: Japanese pavilions, Egyptian and Moorish palaces were erected at enormous expense; one German zoo even built a ruined Gothic castle to display its Barbary sheep upon. Rome Zoo still has a magnificent entrance built in the baroque style and Schönbrunn Zoo in Vienna also has some of its baroque buildings still extant. Little concession was to be made to the varying needs of the animals until the arrival of a fishmonger's son, Carl Hagenbeck, later in the century.

The second reason for the Society zoos' seemingly slow realization of their potential was the same problem that plagues the modern zoo director: finance.

At the time of its conception, Regent's Park Zoo suffered few competitors in the form of established menageries. The only other permanent exhibitions of wild animals in London were at the Tower, shortly to be absorbed by the Zoological Society, and a highly successful menagerie called the Exeter 'Change in the Strand, in fact a travelling menagerie that had established a fixed base in an endeavour to cut the ever-increasing expense of moving from town to town. There were other travelling menageries, too, fairs and amusements for the English public that made regular visits to cities like London. The Haymarket, still part of London's entertainment centre, often had displays of performing animals, jugglers, clowns and acrobats, and Peckham Fair in particular specialized in Wild Beast Shows. As with the Roman Games, these shows were often examples of gross cruelty, ignorance and depravity. Along with the morbidly curious, Peckham Fair also attracted thieves, cut-throats and prostitutes, until in 1827 it became such a public nuisance that it was finally closed down.

The earliest of the Wild Beast Shows that became so popular in England in the eighteenth and nineteenth centuries can be traced back to Pidcock in about 1708. Later prominent exhibitors were Wombwell, Atkins, Bostock and Chipperfield.

The Society zoos had no intention of following such a tradition and in 1825 the prospectus on the proposed Zoological Society of London was circulated. It read:

> It would well become Britain to offer another and a very different series of exhibitions to the population of her metropolis, namely animals brought from every part of the globe to be applied either to some useful purpose, or as objects of scientific research, not of vulgar admiration.

The two men chiefly responsible for promoting the scheme were Sir Stamford Raffles (later the Society's first president) and Sir Humphrey Davy, then one of the nation's leading scientists. When the Society was formed in 1826 it also stated in its charter that it had been created

> For the advancement of Zoology and Animal Physiology and the introduction of new and curious subjects of the Animal kingdom.

These two statements constituted an attitude unprecedented in the history of keeping wild animals in captivity, and although ethically correct, they were to prove difficult to live up to.

£5,000 was set aside for the initial development of the Gardens which, by the time they opened to the public, contained a collection of at least 200 animals, including leopards, kangaroos, a bear, cranes, emus and gannets. But zoological gardens have always been financially top heavy and the money taken at the turnstiles can rapidly decide the fate of a collection.

Today few would deny that the Zoological Society of London has, despite numerous mistakes, largely lived up to the expectations of its founders; frequently it has exceeded them and doubtless it will continue to do so. Yet in retrospect some of the financial backers of the new zoological gardens must have had cause to be reticent about its future. The public, to whom they looked for patronage and financial support, could be expected to be curious at the onset, but would their interest be short-lived? The abomination of Peckham Fair had been extinguished, but business ventures of equally dubious character were still doing big business throughout Europe. Closer to home was the Exeter 'Change.

Originally business premises, the building consisted of three floors, with the animals all housed upstairs in cages that were vulgarly painted to represent jungle scenes, supposedly relevant to their place of origin. A more critical observer called them 'disgusting receptacles'. Eventually the enterprise was managed by Edward Cross, who did little to improve the standards of animal husbandry, but apparently had rare ability as a promoter. Visitors flooded to the 'Change in such numbers that at one stage in 1826 it seriously threatened the survival of the newly founded Zoological Society of London. It is strange to think that this book, written nearly 150 years after the closing of the Exeter 'Change, is about exactly the same problem—that of insanitary, degrading and unprincipled menageries taking the entrance money that should rightfully belong to the more serious zoological ventures of the country.

The death of an elephant named Chunee eventually helped to destroy the 'greatness' of the 'Change. One night the animal became uncontrollable, possibly it was in must. A file of soldiers was summoned and they proceeded to fire 152 bullets into the great beast. The animal was well known by Londoners of the day and they felt affection for it. Many were outraged by its ignoble death and this bad publicity contributed to a decline in the menagerie's popularity. Two years after the incident the 'Change was demolished.

Not only did this type of menagerie exist when the Zoological Society was founded, but public opinion remained uncritical. Richard Martin tried to gain Parliamentary support for a number of Bills designed to prohibit bull-baiting and bear-baiting, but neither was declared illegal until 1835; after all, Parliament had only recently abolished the trade in human slaves (1807), so it was hardly ready to be badgered into passing laws to protect animal life!

'Humanity' Martin, a distinguished lawyer and voluble Member of Parliament for Galway, was an intriguing character and a constant thorn in the side of Sir Robert Peel with his compassionate appeals on behalf of animals. He prompted a contemporary to say of him that he 'appeared to value human life at a lower estimate than the life of a dog or an ass'. But Martin was not a sentimentalist; surprisingly he was never an antagonist of field sports, being of the opinion that they were noble pastimes for gentlemen of refinement. However, he was deeply concerned by the popularity of baiting and its effects on the spectators, whom he considered could only be corrupted by it. At one point he urged that a committee be formed to consider how such sports affected popular public morality. In 1825, whilst Martin continued to champion his cause in both Parliament and the courts of law, two more bulls were baited to death in the vicinity of London, and in that same summer George Wombwell announced that he was to stage the first lion-bait in England since the reign of James I.

Wombwell's Menagerie—it actually comprised three separate shows—was one of the most famous of all England's travelling menageries. Wombwell was a shrewd promoter and showman whose virtuosity and sheer audacity defies imagination. Once, when one of his fairground elephants died suddenly, he had all his showground posters altered to read THE ONLY DEAD

ELEPHANT IN THE FAIR and succeeded in drawing hosts of people from a nearby rival whose elephant was remarkably alive and well!

The lion-bait was staged in Warwick, where a somewhat bewildered and docile five-year-old lion, named Nero, was placed in a barred cage within a makeshift amphitheatre of circus wagons and warehouse buildings. The dogs were set upon him three at a time, yet in spite of reports that the lion had been starved into a state of fury, not once did Nero attempt to bite his tormentors, merely cuffing them aside as they closed in on him. Roaring loudly in pain from a succession of bites to the nose and dewlap, the frightened lion was reprieved when the master showman stopped the contest, much to the disgust of the onlookers, some of whom had paid as much as two guineas for a seat. Undeterred, Wombwell staged a second lion-bait later the same week, but the resulting fiasco was little different from the previous one. Most of the national newspapers reported on the notorious Warwick lion-bait and *The Times* devoted two columns to the affair, damning the torture of a noble beast for mercenary profit, and expressing disgust at 'the supineness of the magistracy at Warwick'. Yet, incredibly, when this unprincipled rogue Wombwell died in 1850, *The Times* stated that 'No-one probably did more to bring forward the study of natural history among the masses'!

So these were the attitudes current when the Zoological Society launched their enterprise: on the one hand, an erudite group of men anxious to further the study of natural history and other new departures in the sciences; on the other, a group of avaricious showmen eager to exploit wildlife with total disregard for possible consequences.

The situation is not so dissimilar today, few laymen realizing that the basic distinction between a zoological collection and a menagerie is one of intent, so that on occasions an extravagant menagerie can still pose as something more significant.

Manchester was the first major collection to be established as a commercial enterprise, in 1836—the only other notable public collections being London, Dublin and Bristol. Founded by a restaurant-owner, John Jennison from Stockport, Belle Vue was originally a coaching-inn on the road to Hyde, which at that time was outside the city in pleasant country surroundings. Jennison's collection was initially small. He was not a naturalist but an industrious businessman whose other interests included an ice-storage unit, a brewery, printing-works and gas-works. He even built his own electric generating plant, thus providing Belle Vue with electricity long before Manchester Corporation had its own supply! Gradually the collection was enlarged; many animals were purchased from a nearby menagerie in Salford which had closed down. But Belle Vue did not develop along planned lines. Like so many zoos, and like Topsy, 'it just growed'. This is not meant to imply that a commercial zoo does not have equal incentive to care for its animals, in fact many of them have made valuable contributions to the science of zoo biology. Belle Vue was one of the first zoos to realize that the practice of keeping animals in over-heated quarters was ill-advised, so a heated yet well-ventilated monkey-house incorporating outside cages was built. At a time when most Society zoos were suffocating their specimens, Belle Vue pioneered ideas of acclimatizing animals to the changeable British climate, even allowing their chimpanzees outside in the winter sunshine.

Since those early days Belle Vue has come a long way. It has a reasonable breeding record, is particularly successful at keeping reptiles, and has a number of modern animal houses. Yet, in spite of this, it is primarily a huge entertainment centre with a ballroom, concert hall and large amusement park, frequently the site of pop concerts, sporting events and circuses. A zoological collection wedged between these distractions is sadly misplaced and aesthetic appreciation of wildlife a virtual impossibility amidst the hubbub of the fun-fair.

Imitation is said to be the sincerest form of flattery, it might also be argued that imitation obviates the need to think for oneself. Nowhere is this clearer than in the zoo world. A number of zoological collections, encouraged by the success of London, modelled themselves on the Society zoos. Amsterdam's Natura Artis Magistra is a typical example, and the many fun-fair zoos now checkered throughout Britain are carbon copies of the first commercial zoo, Belle Vue. Unfortunately, as any lethargic student would vouch, if you copy another's work, you are liable to duplicate his mistakes. In many instances this is exactly what happened in the infancy of the modern zoo. Architectural individuality was to be seen only in the exteriors of the buildings and the elaborate entrances of the zoos. Within, they remained singularly monotonous, each animal house reflecting the limited number of restraint techniques that has proved operable. The carnivore house in Berlin Zoo was very like its counterpart in London. It is not unnatural that London was so constantly imitated, the Society pioneered several types of exhibit never before seen in public zoos. Amongst these was the first reptile display in a converted room in 1849, an improvised exhibit which proved so successful that later, in 1883, a proper reptile-house was built. Modern reptile houses the world over owe their origins to these first endeavours as do aquariums, for London's aquarium, built in 1853, was also the first in the world. But mimicry is not progressive, new ideas in presentation and animal husbandry were desperately needed. The visionary who finally produced these new concepts was Carl Hagenbeck.

### *The revolutionary ideas of Carl Hagenbeck*

Hagenbeck was without question the greatest influence of all on the development of the modern zoological park, yet the realization of his genius was the result of chance. Hagenbeck was not a man of great education. In his book, *Beasts and Men*, he confesses, 'My early education was somewhat meagre'. In fact he attended school for little more than three or four months a year. His father, a fishmonger, had been diverted from his normal business interests into modest activities in animal-dealing and this, subsequently, diverted his son's attention from his studies to caring for the growing menagerie. At 15 the young Hagenbeck finished with school and took over the management of the animal-dealing, whilst his father returned full-time to his fishmongering. For one so young, Carl was shrewd; he quickly realized the enormous potential of the venture, and under his direction the business expanded. By the age of 20, he had virtually all the leading animal collectors working under contract to him alone, as well as a number of his own men employed in the field.

Zoos and circuses were sprouting throughout Europe and there were few choices other than to trade with Hagenbeck. Despite the huge caravans of animals captured by the Hagenbeck teams, for some years it was impossible to keep pace with the demands for even more wild animals. It is depressing to consider the numbers of animals that must have died either on capture, in transit, or later, when finally delivered to zoos that lacked optimum living conditions for wild stock. Hagenbeck, however, did take great pains to ensure the well-being of his stock, frequently accompanying freshly caught animals on the long, arduous journeys to Europe. The size of these living cargoes was truly astonishing. The largest consignment, which arrived from Nubia in 1870, contained at least 60 carnivores, 14 giraffes, 26 ostriches, 5 elephants, as well as buffaloes, rhinoceroses, gazelles and many smaller species. Today fast air and sea travel is such a feature of everyday life that it is hard for us to appreciate the time taken on a journey from Nubia to Germany. We could cover such a distance in hours; Hagenbeck, laden with cages of wild animals, took many months. With such limited choices of transport available it is possible that a period of 18 months to three years might have passed from the actual capture of an animal to its eventual introduction to its zoo home.

The drugs and tranquillizers now used in animal capture were obviously not available to the nineteenth-century collector and, apart from the conventional methods of trapping and netting, the usual way of catching the larger game animals was to run them down on horseback until they collapsed from exhaustion. The alternative was to shoot the mothers and collect the offspring —often considered a more expedient solution even in this century. If you add to the fearful experience of capture, the horrific sensation of being confined to a crate or cage for months on end, bumped and jarred over rough terrain, tossed and pitched across unseen oceans, to arrive in a strange, cold and completely alien environment, only then is it possible to comprehend the suffering and anxiety caused, and the resulting mortality rate. Stress is still one of the major killers when wild animals are captured; in Hagenbeck's day its effects must have been terrible.

Yet, if Hagenbeck and other collectors raped the wildlife of the tropical regions for profit, the effects of their depredations were infinitesimal compared with the irreparable inroads made on wildlife populations by the farmers and sportsmen as they brought 'civilization' to the remoter areas of the world.

So successful was Hagenbeck's animal-dealing business that it began to feel the effects of a recession in the purchasing of stocks by zoos and circuses. Trade remained good, but supply soon exceeded demand and the young director was forced to establish larger, better equipped sites to cater for animals remaining in his care for increasingly longer periods. His animals became far less transitory creatures, so that more permanent relationships were created; thus his considerable experience with wildlife was given the perfect opportunity to flourish. Day-to-day observations at his animal station formed the basis of his many theories and experiments that were to revolutionize the whole concept of the zoological park, and circus.

Hagenbeck was greatly disturbed at the barbarous methods used to train animals for the circus ring, like beating into submission or burning with hot irons, feeling that they were neither acceptable nor necessary in order to get animals to perform tricks. Convinced that gentleness and sympathy would achieve far more than tyrannical cruelty, he proceeded to put his theories into practice with incredible success. His work on animal training has never been bettered, nor is it likely to be. Most reputable modern trainers work on the lines laid down by Hagenbeck.

Whilst Hagenbeck did more than anyone to improve the lot of the hapless inmate of the circus, today the moral question remains. Is it in any way defensible to keep animals in cramped circus wagons for most of their life, in order that twice a day they can be admitted to an arena for the sole purpose of performing a series of ludicrous tricks before an audience?

In Carl Hagenbeck's day the main objection to circuses was that of cruelty, a factor he was largely able to eliminate (although there is no doubt that elements of cruelty still exist under many big tops). In this age the criterion is fundamentally a question of whether or not animal acts in a circus have any value to a society that badly needs a greater appreciation of wildlife and environment. Children grow into adults carrying the colours and impressions of childhood like sacred possessions. These impressions, whether true or false, once fixed in the adult mind become ideas that are extremely difficult to alter and can form definite barriers between fact and fiction. For example, many people find the appearance of chimpanzee minus human clothes and other apparel far stranger than one dressed as a clown, milkman or businessman, and it is not uncommon to find that seemingly mature and intelligent citizens will extol an animal's qualities on the basis of the number of tricks it can perform. In this sense, the circus only exploits the human tendency to anthropomorphize.

But of far greater consequence from the zoological viewpoint were Hagenbeck's revolutionary studies of the acclimatization of wild animals. His attention was originally drawn to the problem upon seeing a chimpanzee in England playing in the snow, and later when he came across a menagerie in Westphalia that also allowed its primates outside in the winter months. Fascinated, he began to ponder on how advisable it might be to expose tropical species to the extremes of European winters. The advantages, if it were feasible, were obvious. Access to fresh air was of first importance, but it would also be possible to build larger enclosures, giving the animals far greater freedom of movement. The prospect was very exciting. Clearly, he argued, acclimatization to the weather of a different continent is just one aspect of a larger and more complete acclimatization. A newly caught animal has to adjust to an artificial environment; any number of unfamiliar sights and sounds and in all probability a substitute diet. So, why not the weather? Cautious at first—after all, these ideas were contrary to all the established beliefs of the day—Hagenbeck's first experiment was the result of unusual negligence on his behalf. Awoken in the early hours one winter morning by the call of a bird, he remembered to his horror that he had neglected to shut a Sarus crane in its shelter the night before. Hagenbeck's account of the incident is as follows:

> I hastened out to find the hoar frost lying on the ground, and expecting to see the crane frozen through with the cold. But to my astonishment I found him in the most boisterous health; and when I came up, expecting to see him on the point of death, he came dancing and fluttering round to greet me, filling the air with his loud cries. Seeing that he did not appear to have suffered in any way from the cold, I arranged in one corner of the enclosure a sort of recess with plenty of straw where he could obtain shelter from the cold. But never once during the snowstorms, wind and rain of that long winter did he make any use of this recess. He maintained his health in as perfect condition as though he were in his own tropical climate. I date from this occurrence the inauguration of my settled custom of giving wild animals access to the open air to the greatest possible extent.

Hagenbeck now embarked on the series of experiments on acclimatization that were to occupy a large share of his attention for the remaining years of his life. He found that many species were capable of making a gradual adjustment to low temperatures, and over the years he experimented with a variety of species including cranes, flamingos, ostriches, lions, leopards, tigers, giraffes, zebras and antelopes. Generally, the results were encouraging. Acclimatized animals often grew a thicker pelage at the onset of the colder weather and so long as they were allowed access to shelter or, in some cases, partially-heated quarters, they suffered no ill-effects. The long-term results of the experiments proved conclusively that acclimatization was not so difficult as it had at first been believed. Providing exposure to lower temperatures was controlled so that the animal had a reasonable amount of time to adjust—that is to say by importing newly captured specimens in springtime, not late in the year—the effectiveness of the husbandry was to be seen in the improved health and longevity of the subjects. A hundred years ago the life expectancy of many zoo animals could be measured in months more often than years, but today, thanks in no small part to the work of Hagenbeck, the longevity of captive specimens can greatly exceed that of their wild counterparts.

Zoo directors, stimulated by the success of his experiments, took to experimenting themselves. On his estate in the Crimea, Frederick von Falz-Fein (the founder of what is now Askania Nova), obviously influenced by the work of the great German, took to keeping all his animals out in the open, as did the eleventh Duke of Bedford in his great park at Woburn. In a period of zoo history dogged with misconceptions and uncertainty, Carl Hagenbeck was emerging as a beacon to guide those who wished a better understanding of wildlife.

Up until this point Hagenbeck had involved himself in animal-dealing, catching, training and acclimatization; in each case his successes were dazzling displays of ingenuity and an unequalled understanding of wild animals. The firm was vast, comprising a large circus, animal-catching teams, and a costly base at the *Neuer Pferdemarkt* to house the valuable stocks of wild animals which awaited buyers. To establish such authority in areas never before explored would content most men, but not Hagenbeck. The best was yet to come. For many years he had travelled across the world on business trips to potential clients—zoos and private estates—making a quantitative appraisal of their relative merits or failures. So it is not surprising that such a man began to dream—to dream of a zoological garden of a type never before seen nor indeed dared.

Having proved that whips, clubs and hot irons were unnecessary attendants to animal training, was it not possible to prove that pits, chains, iron bars and even fences could be dispensed with? Again Hagenbeck began a series of remarkable experiments. First he suspended a dead pigeon from the branch of a tree and in turn he released lions, tigers and leopards into the enclosure to see if they could bring it down. In this way he soon found that without a run none of them could accomplish a vertical jump of more than 10 feet. The next stage was more difficult. He now needed to carry out a long jump test. Yet again, Hagenbeck was in a unique position, for all of the animals in the Hagenbeck circus, if not actually trained by him, had been trained according to his rules. As a result they were ideally suited to phase two of his experiments, being far fitter and more athletic than the average zoo specimens yet equally tractable and tolerant of his persistent demands. Concentrating primarily on carnivores, his favourite beasts, he found that on average they could cover 10 feet from a standing jump. He estimated that with a run they could not leap more than between 15 to 20 feet. The experiments continued and information on the leaping ability of many species was gathered from every available source.

Even while involved in these experiments, Hagenbeck was searching for a site for his dream zoo. Unable to find one in his native town of Hamburg, in 1902 he was fortunate to find some land at Stellingen, a short distance from the town. Within two days he had purchased the property and a day later he had bought two neighbouring estates. Only Hagenbeck's words can describe his excitement:

> Now at last I was in a position to carry out my long-nursed project of founding a Zoological Park of a totally different kind from anything that had been before attempted. I desired, above all things, to give the animals the maximum amount of liberty. I wished to exhibit them not as captives, confined within narrow spaces, and looked at between bars, but as free to wander from place to place within as large limits as possible, and with no bars to obstruct the view and serve as a reminder of captivity. I wished also to show what could be accomplished in the way of acclimatisation. I desired to refute the prevailing

notion that luxurious and expensive houses with complicated heating apparatus were necessary for keeping wild animals alive and healthy. I hoped to show that far better results could be obtained when they were kept in the fresh air and allowed to grow accustomed to the climate. I wished my new park to be a great and enduring example of the benefits that can be wrought by giving the animals as much freedom and placing them in as natural an environment as possible.

The concept alone was brilliant. Not only did Hagenbeck appreciate that biological conditions for wild animals could be created within an artificial environment, therefore dispensing with any feeling of being confined or restricted, but he also understood that it was equally important for the human observer to lose notions of 'captivity' in order to assimilate and glean the aesthetic pleasure of being part of wildlife. It was to take five years of hard work, and Hagenbeck's fruitful mind, coupled with the design ability of the Swiss architect, Eggenschwiler, to transform the site, and the dream into reality. On 7th May 1907 the gates were at last opened on the crowning triumph of Hagenbeck's career and the first visitors were astounded and awed by what they saw. Enormous artificial mountains had been raised from the virgin land to accommodate the wild sheep, chamois and ibex; large glens contained lions and tigers, separated from the public by carefully concealed moats. Dramatic panoramas, using not only concrete mountains and hidden moats, but also artfully placed trees and shrubs, showed animals of one continent, both carnivore and herbivore, one against the other. It was not only the most spectacular zoo in the world, it was the most advanced; Hagenbeck had transformed the keeping of wild animals in captivity into both a science and an art.

In 1911 Rome Zoo was opened, also built by Hagenbeck; and at least four others were to be established on the lines laid down by the master. His influence on zoological park development is irrefutable. The first barless zoo was not just an inspired piece of showmanship, it gave others a clear path to follow; and for the first time in its inglorious history, the menagerie had been pulled out of its excrement to reveal an institution worthy of civilized man.

In 1943 an air-raid lasting approximately 90 minutes accounted for the destruction of 80 per cent of Stellingen Zoo, the deaths of eight people and 450 animals. Fortunately Hagenbeck's grandson immediately took charge of the reconstruction of the zoo, making improvements to the original concept wherever possible, and the task was complete in 1960. The famous Hagenbeck circus is now a thing of the past, although a small animal show is given each day in the building where Carl Hagenbeck worked on his theories of training animals without cruelty. Stellingen itself remains the finest example of the use of moated enclosures in zoological parks.

Among those impressed by the success of Hagenbeck's *Tierparke* was a Mr T. H. Gillespie, who put considerable effort into forming the Zoological Society of Scotland as a result of what he had seen at Stellingen. The Society was formed in 1909—two years after the opening of Stellingen—but experienced considerable difficulty in raising capital, so it was not until three years later that a site was purchased for the proposed zoological garden. This site, on the Corstorphine Hill House estate—now Edinburgh Zoo—was vigorously attacked by the planners and builders in an effort to open to the public as soon as was possible. Much was built on Hagenbeck's principles, in particular the two most important exhibits for lions and polar bears, and ditches and large enclosures were a leading feature of the new zoo.

### *Peter Chalmers-Mitchell and Whipsnade*

Whilst Mr Gillespie had been inspired to form a zoological society, another great zoo man, Sir Peter Chalmers-Mitchell (Secretary of the Zoological Society of London 1903–35), saw Stellingen as a substantial piece of evidence that supported the theories that he himself had nursed for many years. Like Hagenbeck, Mitchell had also undertaken acclimatization experiments and firmly believed that wildlife, particularly herd animals, would benefit from fresh air and greater freedom of movement. Upon taking office at Regent's Park, Mitchell carefully scrutinized the records of deaths within the zoo covering the period from 1870 until the turn of the century. By comparing the longevity of the animals at the zoo with the estimated life-span in the wild state, he concluded that the average mortality rate was far in excess of that which should have been expected. Captive animals had the benefits of a constantly available supply of food and water, whereas in the wild they experienced long periods when neither could be found in quantity; in captivity there were no competitors or predators. The one fact that came to light from his studies was that in the few instances where the zoo had set up a longevity record with a species, invariably the animal in question had had access to fresh air. Mitchell's summing up of his work was as follows:

> The most fatal type of housing for any mammal or bird is confinement in the interior of a warmed house without free access to the open air. The conditions are very complex and I do not suggest that the provision of heat is in itself an evil. The ideal throughout the period on which my statistics were based, and which still maintains an evil existence in the minds of a majority of those who have to do with living animals, is that in the first place they have to be kept warm. Warmth having been secured, the more advanced persons have consented to, or even urged, the advantage of ventilation, moistening of the air, and so forth. The idea, however, is wrong. The first requisite is free access to the open air, the rest is light, space and cleanliness; these things having been secured, any form of heating that may be thought advisable may be added, in so far as it does not in any way interfere with the primary considerations.

Comparison with Hagenbeck's findings is inevitable. One of Mitchell's most important experiments was also one of his earliest. In 1903 he placed a number of baboons in open cages and eliminated all artificial heating. Careful observation of the animals showed conclusively that the benefits distinctly outweighed any disadvantages experienced and when the great Hagenbeck visited London and gave his unqualified seal of approval to his work, Mitchell must have been delighted.

Although many schemes for the advancement of zoological park techniques were nurtured in the mind of Chalmers-Mitchell, it was to be many years before they were to be realized. The famous Monkey Hill at Regent's Park was the ultimate result of his early experiments with baboons, yet not until 1925 was he able to persuade the Society to establish such an important exhibit. The same applies to his greatest achievement, the Whipsnade experiment, first suggested by Mitchell as early as 1903, and repeatedly rejected by the Council. It is regrettable that the attitudes to conservation of many zoological councils have so often hampered the progress of gifted men. In this light Hagenbeck can be seen as fortunate not to have been hindered by the bureaucracy of committee meetings. But in 1926 Mitchell finally obtained the undivided attention of the Council, the majority of members at last accepting that they were dealing with a man of remarkable ability and foresight in zoological matters. The original idea had been simply to provide a country rest centre for animals, but now the aims had been expanded and the Whipsnade scheme became far more ambitious. Mitchell hoped that by providing animals with far more natural surroundings than was possible in the confined spaces of Regent's Park, natural groups and herds could be established, resulting in an improvement in breeding potential, and giving many species the opportunity to display normal social behaviour. Such work was becoming increasingly important; attention was being focused on the depressing fact that many species of wild animals, once abundant, had been so reduced in numbers that their survival looked exceedingly precarious. A zoologist of Mitchell's calibre would have been acutely aware of comparatively recent losses and, from the conservationist's point of view, missed opportunities. The bluebuck (*Hippotragus leucophaeus*) was exterminated by 1800, the great auk (*Pinguinus impennis*) in 1844, Stellers sea cow (*Hydrodamalis stelleri*) by 1854 and the last quagga (*Equus quagga*) died in Amsterdam Zoo on 12th August 1882. In the New World our American cousins had managed to exterminate the passenger pigeon (*Ectopistes migratorius*), once believed to be one of the world's most prolific bird species, and had relentlessly persecuted the plains bison (*Bison b. bison*) until the population had been reduced from an estimated 60 million to a few hundred by 1889.

So, apart from its educational function, the zoo was beginning to be seen as important for the preservation of rare animal species. Mitchell was not slow to appreciate this fact; valuable contributions had already been made by men like Falz-Fein and the 11th Duke of Bedford when they gathered together some of the few remaining Przewalski's horse (*Equus przewalskii*) and established breeding herds on their estates. The Duke of Bedford, who was President of the Zoological Society at the time of Mitchell's appointment, had a large collection of exotic animals on his lands at Woburn, including a number of rarities. It was he who commissioned Carl Hagenbeck to go to Mongolia and capture specimens of Przewalski's wild horse from the few herds that remained at that time. Over 50 foals were captured, several of which were purchased by continental zoos, although the pick of the lot went to Falz-Fein at Askania Nova and the Duke of Bedford at Woburn. The operation could hardly be classed as an exercise in conservation as the capture of these specimens involved the shooting of the foals' mothers and, on occasions, the herd stallions, but such methods were considered usual at the time. These catching techniques showed an irresponsibility that is still commonplace with some animal dealers. In the case of the Przewalski, if the animals had not bred in captivity, the species could have been pushed to an early extinction by the combined efforts of those who desired to save them. Fortunately this is not what happened.

It is possible that a few wild horses still exist in the wild, although certainly less than 50 (some authorities consider even this unlikely), but the captive population in January 1974 was 219. Of these, 93 are stallions, and it must be remembered that the remaining 126 mares will not all be of breeding age or condition. Some experts have already stated that the reproduction of the world Przewalski herd is not so high as it might be and this is largely due to the distribution of the animals—124 are contained in 10 major collections whilst the remainder are kept in 42 collections. There is strong evidence to suggest that there is less chance of breeding success with these animals when they are kept in small numbers as opposed to large herds. So there is still a possibility that we may lose the last of the wild horse, an animal listed as one of the world's 25 most endangered species, if we do not act responsibly and quickly. Yet the fact still remains that the species has survived until now solely because of the early efforts of Falz-Fein and the Duke of Bedford.

Woburn is even more famous for its magnificent herd of Père David's deer (*Elaphurus davidianus*). These are the strangest of all the true deer and the Chinese name for them is *Sse-pu-hsiang*, which literally translated means 'four-in-one', as the Chinese felt that this deer possessed the characteristics of the deer, the ox, the reindeer and the donkey. It is certainly an odd-looking animal with an unusually long tail and antlers that are quite unique among the deer family, appearing to be back to front at first sight. The hooves are large, with long pseudo-claws, resembling those of the reindeer and making a loud clicking noise as the deer walks. The species ceased to exist in the wild about 2,000 years ago, but survived in the Imperial Hunting Park near Peking until its scientific discovery by a young French Jesuit missionary, Père Armand David. David, an enthusiastic zoologist, recognized a species hitherto unknown to the sciences of Europe and managed to procure two skins which he sent to the zoologist Milne-Edwards. In 1886 Milne-Edwards published a paper on the species and shortly afterwards many European diplomats began to make overtures to the Emperor for living specimens for their animal collections. France, Germany and England were the main beneficiaries and soon there were approximately 24 of the animals settled in Europe. The 11th Duke of Bedford was concerned at the way the deer had been dispersed and determined to obtain as many as possible. By the end of the last century he had acquired 16.

In 1895 a flood breached the 45-mile wall that enclosed the main herd of Père David deer at the Imperial Hunting Park and all but a handful escaped, only to be killed by the starving Chinese peasants. Five years later, during the Boxer Rebellion, the rest were slaughtered, leaving the Woburn collection the only herd in the world. By 1922 the Woburn herd had increased to 64 and plans for the creation of new herds were being made. Today, thanks entirely to the intervention of the 11th Duke of Bedford, the world population is over 400.

The value of such work was, and still is, undeniable, so from the start Chalmers-Mitchell regarded wildlife preservation as one of the main functions of the Whipsnade scheme. Significantly he also proposed that the new Country Park zoological collection should increase the number of specimens whilst limiting the number of species kept. In effect, he was advocating the advantages of specialization over the conventional zoo policy of showing the maximum number of species with a few individuals of each.

In 1931 Sir Peter Chalmers-Mitchell had the great satisfaction of knowing that, after years of campaigning, he was largely responsible for launching an entirely different kind of wild animal collection, a collection that is today firmly established as one of the greatest in the world. At its foundation it was believed that Whipsnade could not possibly be a financial success and would need continual financial support from the parent collection at Regent's Park. However, it soon began to break even and eventually made substantial annual profits which were ploughed back into the Society's research programmes. In the limited space available within a short chapter it would be quite impossible to do justice to Whipsnade's many achievements, but it is worth mentioning its major contribution to our present-day knowledge of the optimum conditions needed for maintaining animals successfully in captivity. It has been clearly shown that many species that appear to do well under captive conditions do not breed. Therefore, the requirements for captive breeding are often far more critical than those for survival. Experience has also shown that these criteria vary from one species to another, so that the optimum conditions for one type of species will prove totally unsuitable for another; even closely related animals can show extreme differences in their biological requirements. Within the parkland of the new zoological park large areas were made available to the animals so that it became possible for many species to establish definite territories. This, in particular, has been an important factor with the high breeding success of the famous herds of ruminants, including the many species of deer in the collection. Rare and difficult species bred in the park include musk ox (*Ovibos moschatus*), moose (*Alces alces*), spectacled bear (*Tremarctos ornatus*), European bison (*Bison bonasus*), snow leopard (*Panthera uncia*), Indian rhinoceros (*Rhinoceros unicornis*), Przewalski wild horse (*Equus przewalskii*), Père David's deer (*Elaphurus davidianus*), cheetah (*Acinonyx jubatus*), Manchurian crane (*Grus japonensis*) and Asiatic wild ass (*Equus hemionus*).

But it would be wrong to consider Whipsnade an unqualified success. It has passed through periods of financial and administrative difficulties, although its present redevelopment plans look very promising. It has also witnessed the meteoric rise of strong competition with the rapid escalation of zoo developments throughout Britain.

The real significance of Whipsnade, Sir Peter Chalmers-Mitchell, Hagenbeck and his *Tierparke* at Stellingen lies in the fact that they heralded a new era for the zoo. It would be erroneous to believe that theirs was the only contribution to this new age of zoo biology, but theirs was so overwhelmingly important it has been used as the best example for this brief account. Because of the phenomenal trail-blazing ability of some of these early major zoological parks and their directors, we have learnt more about the physiology and the psychology of wild animals in the last 50 years than civilized man had known right up to the end of the nineteenth century.

But this brief zoo history should not be considered comprehensive, nor is it intended to be so. We have paid scant attention to certain historical periods, whereas in some instances what must seem an unreasonable amount of space has been devoted to certain zoo figures. Hagenbeck was certainly not the greatest animal lover of all time, he can hardly be considered a great conservationist, yet he did inspire a form of wildlife exhibition that greatly increased human involvement with animals, without imposing detrimental restrictions upon the animals themselves. It was a big step forward in the art of zoo display.

The story of precisely who owned wild animals in the past still seems largely irrelevant. On the other hand, why they were owned is highly significant, for, as we have already stated, the very best qualities of a zoo result from a specific attitude of mind. Whilst in this century there are undeniably men who might be compared with the altruistic Akbar—men like Sir Peter Scott, Phillip Wayre and Gerald Durrell, who view the zoo as a vehicle for wildlife preservation—they are greatly outnumbered by present day Wombwells. The British zoo has in fact failed to establish a positive identity or make a declaration of intent, so how can the average visitor know what to expect from his local zoo if the zoo establishment as a whole is unsure of itself? Some commercial enterprises have been quite honest about their activities, and are not ashamed to admit that they operate solely for monetary gain; others pretend to be of a more serious nature. However, zoos like Jersey, Marwell, Port Lympne, Twycross and Norfolk make no bones about the fact that they admit visitors because they need the money to finance their breeding programmes, and that their main objective is wildlife preservation, coupled hopefully with an improvement in the layman's appreciation of the natural world.

Unfortunately, the situation is not so straightforward. In May 1972 the first World Conference on the Planned Breeding of Endangered Species in Captivity was held in Jersey. The meeting was sponsored by the Fauna Preservation Society and the Jersey Wildlife Preservation Trust. This conference was a milestone in zoo history because it was the first real opportunity for zoo directors from all parts of the world to get together and exchange ideas, to consider their responsibilities and decide on a definite course of action. The main theme of the conference was an appeal for greater co-operation between zoos, and an acceptance of their responsibility to become 'wildlife banks', acting as breeding centres with a view to returning captive-bred endangered species to their natural habitats.

The second World Conference was held in London in July 1976 and followed the same theme as the first, save for the closing address. This was delivered by Lord Zuckerman, President of the Fauna Preservation Society and long-time Secretary of the Zoological Society of London. Despite the fact that human survival and wildlife conservation are considered one and the same by conservationists the world over, Zuckerman found it necessary to set wildlife preservation against other priorities like combating starvation. The implication that the human race had a choice between starvation and wildlife preservation was misleading to say the least; from so respected a figure in wildlife circles it must be considered regrettable.

'Who said we have to return rare animals bred in captivity to the wild?' asked Lord Zuckerman. 'Let us ask how many people's lives are worth sacrificing for the preservation of a species of animal.'

It is likely that this ill-founded address harmed the reputation of the Society and the endeavours of those zoos actively engaged in attempts at wildlife preservation. Certainly dismay and embarrassment appear to have been the prevailing reactions among dedicated zoo people who had come from all over the world to attend the conference, although the episode undoubtedly caused considerable amusement amongst the commercial zoos.

If Zuckerman was correct in his assessment of the zoo, then he produced the most positive case against the zoo's continued existence. If he was wrong, and we believe he was, then he contributed greatly to the confusion that has existed since the dawning of the zoo concept. We are left with a puzzlement. What is the bounty of our age: a peepshow, or a bank for research, education and conservation?

Chapter 2

# Why zoos?

Two governing influences make a wild animal collection what it is, good or bad: the visiting public and the zoo's governing body.

Is there such a thing as a typical zoo visitor? From our own experiences the answer seems to be no. There are those who see a day at the zoo as an outing filled with elephant and camel rides, beer and sandwiches, fun-fairs, ice-cream, candy-floss and the animals' feeding-time. Feeding-time is the highlight of their day; lions tearing bloody flesh, dolphins and sea-lions performing their daily acrobatic ritual in return for fish, the primates sorting through fruit bowls for their favourite delicacies. To this fairly innocent group the zoo is simply an ideal place to take the children, especially in the long summer holidays.

There are those who visit zoos for a giggle, in much the same way as our Victorian ancestors visited Bedlam in order to laugh at the antics of the unfortunate inmates. This group find everything from the chimpanzees' friendly countenance to an elephant defecating cause for riotous laughter, and if such stimulants are not present in abundance they will resort to teasing the animals in order to generate their own brand of entertainment. Monkeys are common prey to this kind of torment as they react very quickly to any form of harassment. Once roused, they rush angrily at the cage front with mouths opened to expose their savage rows of teeth, whilst the tormentor, confidently secure with a solid barrier between him and his victim, slowly increases the level of irritation. Often an animal that is unable to attack an enemy, which is the role assumed by a person who teases animals, will re-direct its aggression towards its companions; in this way arguments can develop within a group that will result in physical injuries and sometimes actually contribute to the disruption of the social cohesion within the group.

Whether perverts are an extension of this latter group or a class of their own is a matter for sociologists, but they exist. Though they only represent approximately one per cent of zoo-goers, they cause problems for a lot of zoo directors. At the saddest level, this group is personified in the dirty mac brigade, solitary individuals who roam the zoo with eyes forever focused on the animals' genitals, hopefully looking for a lascivious mating scene. But generally they cause little trouble. One cannot say the same about the activities of the sadists, however.

Poking eyes is the most elementary of their pursuits. When inspired they will scatter broken glass and nails in the cages and enclosures, or conceal razor blades or poison in morsels of food. Chester, one of the best of Britain's large mainstream zoos (mainstream zoos are those which exhibit a representative collection of vertebrate animals, mammals, birds, reptiles, amphibians and fish), was the victim of such a maniac when in the summer of 1976 several of the big cats were poisoned. Although rare, incidents like this have persisted for a long time and it is difficult to foresee a permanent solution to the problem, save banning the public altogether. This is not so idiotic as it may seem, but it is a debatable point to which we shall return later.

Finally, a lot of people go to zoos because they are genuinely fascinated by animals. This group is represented by the regular zoo visitor, who will often make himself known to the staff and can be surprisingly knowledgeable about the species of animal on show. Unfortunately a large number of them are remarkably uncritical of the conditions the animals are kept in, probably because they are so besotted with the beautiful creatures that they pay little or no attention to the environment in which they are forced to exist. As one leading preservationist recently said to us, 'The world's last surviving wild animal will ultimately be crucified by a bloody sentimentalist!'

This bizarre assortment of types hardly constitute an ideal electorate to decide the future of the zoo, nor do they provide a case for its continued existence.

The other major influence on a zoo's development is the owner, management or council, as the case may be, but often they are partly or even wholly influenced by the demands of the public. This is certainly the case of Britain in the last two decades. To be guided by the demands of a public that, in truth, has no idea of what it wants or should expect from a zoological collection must result in an anarchical state of affairs. If this sounds as if we are placing an unfair amount of blame on the visitors, it is not intended. Uncritical visitors certainly encourage the exploitation of wildlife, but a parent might just as well blame a very young child for breaking a rare antique that should properly have been placed out of harm's way, as we might blame the visitor for the lamentable state of our British collections.

Pandering to the public is to some extent necessary, because every collection relies entirely upon its patronage to remain solvent. In most cases the visitor is the sole economic life-line of the establishment. That this is so is extremely regrettable for several reasons, not the least of which is that often the needs of the public are quite irreconcilable with the requirements of the animals. Yet none of our zoos receive any form of government aid, except London, and the only legitimate way to increase revenue and safeguard it against excessive tax is to register the project as a charity. This has been done with a number of the best collections, including The Wildfowl Trust, The Jersey Wildlife Preservation Trust, The Pheasant Trust, The Otter Trust and the East Midland Zoological Society. Nevertheless, these groups are still primarily dependent on the paying visitor and are obliged to provide facilities for them and even to think in terms of the 'exhibition value' of certain creatures. So let us consider how this almost total dependence on money from visitors can be counter-productive, actually threatening the values that are inherent to zoo philosophy.

To start with, every zoo project must have a good location in what developers call a 'honeypot' area. This is situated either near a city, several large towns, or at a popular seaside resort which is guaranteed to attract a large holiday population. To site a zoo otherwise, in an obscure hamlet miles from well-populated conurbations for example, would be sheer foolhardiness, whatever the motivation behind the project. Having acquired a suitable property and assessed the expected attendance figure, the zoo has then to provide the public facilities: lavatories, tea-room or restaurant, car parks, first-aid posts, seats, litter-bins, informative labels and direction signs. All these are essential when large numbers of people have to be catered for. The gift-shop, now found in nearly all collections, seems to mark the borderline between good taste and blatant vulgarity. People like to purchase small items to mark their visit to the zoo, and guide-books, post-cards, natural history books and wooden or pottery animals are ideal mementos of a day out. Some of the other rubbish sold in these shops however seems remarkably out of place, bearing no relationship to the natural history theme at all. Toy cars, playing-cards decorated with naked girls, kites, footballs, costume jewellery and perfume are only a few items of the irrelevant junk we have seen displayed in several gift-shops across the country.

So what has this got to do with zoo standards? In our view it has a great deal. A visit to a zoo is largely an atmospheric experience which should not be cheapened by trivia. Zoos run these shops to make money, the mark-up on many of these gifts is high (often 100 per cent), but in so doing they should use a little grace and endeavour to retail items that have some significance to a day at the zoo. But, as we said, the gift-shop is a borderline case; not so the other amenities. The profit motive alone is responsible for the introduction of fairground amusements to the zoo grounds. To believe that such extraneous amusements are necessary is to admit that the animals alone are not sufficiently attractive, that the zoo management is out of touch with zoo biological principles and that any declaration of purpose behind the project can be dismissed as no more than public relations propaganda. The world's finest zoological parks do not contain amusement arcades or fun-fairs because they have no need of them, whereas those zoos that do often show inferior standards of animal husbandry. Quite apart from this, the conflict between the appreciation of wild animals and the cruder absorption of the fun-fair results in a weird, unpalatable concoction. Anyone who has visited Belle Vue in Manchester, Chessington in Surrey, or Drayton Manor, Warwickshire, will understand what we mean. On entering the gates, one ponders on which attitude to adopt to tour the place: should one view it as a fun-fair containing wild animals, or as a zoo containing a fun-fair? All of these collections would be improved overnight simply by removing the unnecessary amusements, and there is no reason to suppose that they would suffer a decline in attendance as a result. If anything, they would probably find a resurgence of public support, because each year it seems that more and more people find the fun-fair repugnant.

Not all zoos have amusements, but the ill-advised policy of giving the public what it wants manifests itself in other ways. Cramming as many different species as possible into a limited space is the commonest error and the results of this policy will be discussed in more detail later. But the saddest effect of all is when a really good collection is forced to compete in the commercial rat-race and devise exhibits it would be better without.

One of our favourite collections is Twycross in Warwickshire. Here one will find the most comprehensive collection of primates in Britain, housed in excellent accommodation, with their every need catered for by an energetic and experienced staff. The standards of husbandry compare with those of the best zoos in the world and are far in advance of all but a few in the United Kingdom. Yet in the midst of this marvellous zoo there is an exhibit that is, in our view, totally unsuited to the zoo's style and alien to their highly specialized knowledge—an elephant house. By comparison with elephant houses throughout the country it is a reasonably good building (although the outdoor enclosure is far too small). Under normal circumstances this would not merit severe criticism, but in a zoo renowned for its positive attitudes and its resourcefulness, it seems a negative feature. One female African and one female Indian elephant are kept here. They are well cared for, taken out for regular walks and beyond question they are far better off than most of their relatives in other collections in the country. But what purpose does this exhibit serve? Cross-breeding elephants seems an unlikely eventuality. With two females as breeding-stock it's an indisputable impossibility! Perhaps later on Twycross intends to enlarge the existing building and acquire more of each species; but if they did they would be extremely foolish, as the capital expenditure involved would be far in excess of their resources and could achieve little, save to detract from the valuable work already being done in Twycross.

We know that the management are not entirely happy with the exhibit, so why was it built in the first place? Here we are obliged to draw our own conclusions. Twycross is situated to the east of Birmingham, approximately 20 miles from the city centre, and in easy reach of several large Midland towns. These densely populated areas afford the zoo the prospect of high attendances throughout the summer season. However, in 1977 there were no fewer than six zoos, two safari parks and several bird gardens, all within a 60-mile radius of Twycross, and all competing for the same visitors from the Midlands. Although the Twycross collection is far superior to its local competitors in terms of accommodation, husbandry and technical expertise, these factors are generally missed by the public, so the organization finds itself in the unenviable position of touting for custom. To this end it has become a general, mixed collection, containing lions, tigers, giraffes, camels, tapirs, penguins and sea-lions, simply because this is what the public expect to see in a zoo.

Their breeding record with several of these species, especially the giraffe and tiger, is remarkably good, but we feel it would be regrettable if this general selection of animals was extended to any great degree. Britain just does not need any more general collections, but specialized collections are urgently required and Twycross has the potential to become the leading primate centre not only in the United Kingdom, but possibly in Western Europe.

It is perhaps unfortunate that we have chosen one of our finest zoos as an example of the effects of commercialization, but it is also unfortunate that so admirable an institution is held back in its evolution by public demands and the existence of commercial competition. Twycross should eventually have laboratories, a research centre and facilities for combined studies in the wild and captivity; but until the public show a more selective attitude in their zoo tastes this Utopian dream will long remain unrealized, and the zoo will have to continue to plough back every penny taken at the gates simply to maintain the exceedingly high standards it has set itself.

The popular image of the zoo has much to answer for, because, of about 150 collections in this country, the majority are very similar in content and style. Of the large Society zoos only Chester is, in our view, truly exceptional. One safari park is invariably the same as the next, as are wildlife parks and small or medium-sized zoos. On the whole, bird gardens are usually the most interesting and varied places to visit and basically this is because £1,000 can be put to greater effect in the construction of an aviary than it would be with, say, a tiger enclosure.

Clearly, for as long as zoos themselves perpetuate the common zoo image, the development trends in zoo biology must remain a point of academic discussion instead of reality. And, surprisingly, some of the well-established scientific zoos are as guilty of subscribing to this negative policy as the openly commercial groups. London Zoo is particularly guilty in this respect. Despite their serious-minded declarations of zoological intent, one is hard-pushed to appreciate it when visiting the place. It is highly commercialized in presentation, yet very unimaginative with regard to other factors. Monolithic elephant-houses and giant aviaries may look very fine but they are not necessarily better or worse than other more modest constructions elsewhere.

London is open to attack if for no other reason than that it is the oldest and largest collection in the British Isles and therefore might be expected to give a more definite and forceful lead to other collections. This it has largely failed to do. Consequently it has become the butt of severe criticisms from a variety of sources, and in 1976 several articles were published that openly attacked

1 Lion in Serengeti National Park, Tanzania

2a. Zoo or funfair?
b. Who is entertaining whom?

3 Examples of appalling crating of animals, found by the RSPCA's hostel at Heathrow Airport

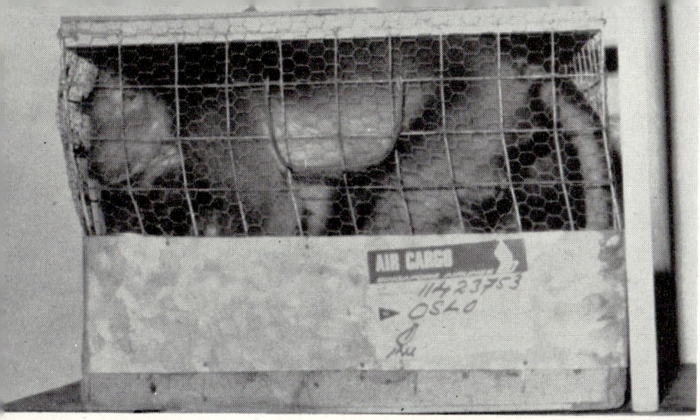


a. The cramped, inadequate crating for this baboon has buckled in transit

b. Two dead tigers and the crate in which they travelled to Britain

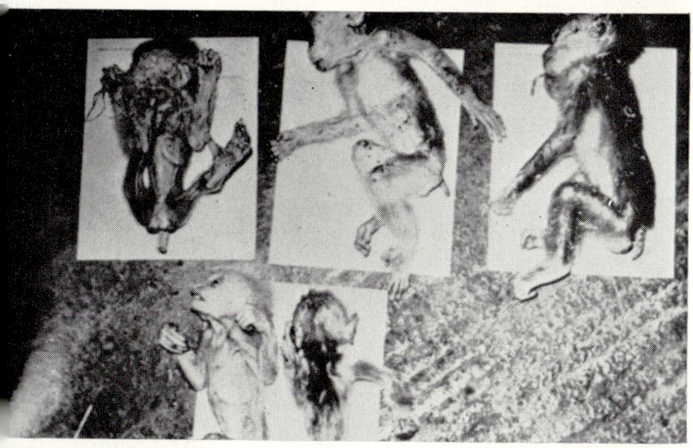

c. These pathetic baby monkeys died in transit, because they were far too young to be exported

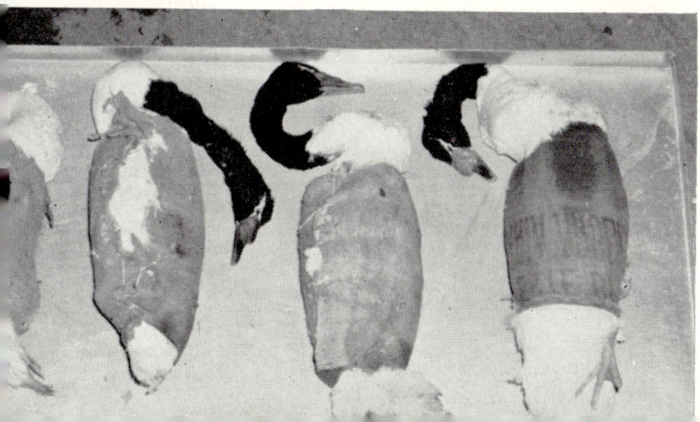

d. Four dead black-neck swans sewn into sacking

4. Transportation

a. Loading a partially tranquillized white rhinoceros into its crate

b. 20,000 feet over the Sahara, a black rhino gets some refreshment

c. Two elephants enjoy a dustbath after a 2,000-mile journey in their crates

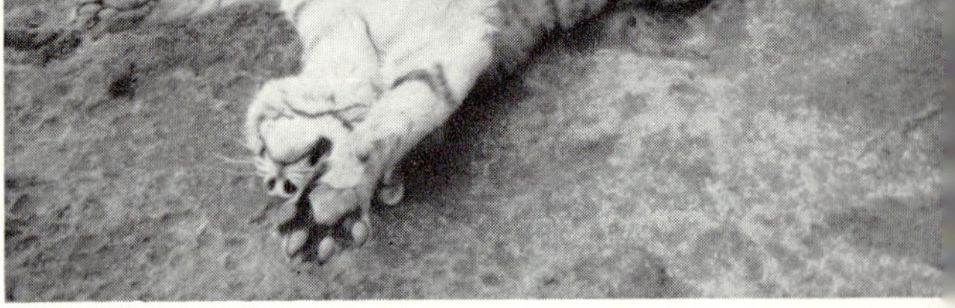

d. A Siberian tiger rolling on its back. This snapshot was taken only two minutes after it had been uncrated

5 Reticulated giraffe in its natural surroundings, Samburu, N. Kenya

6a. This zoo giraffe's feet are overgrown, due to lack of a hard surface to wear down the horn. Note the mud

b. Hyper-aggressive display by a Hamadryas baboon

d. Do elephants have to be chained? Note spiked lower rail, effective but crude and chain marks just above feet

c. This tapir has a skin infection. It has no access to a pool

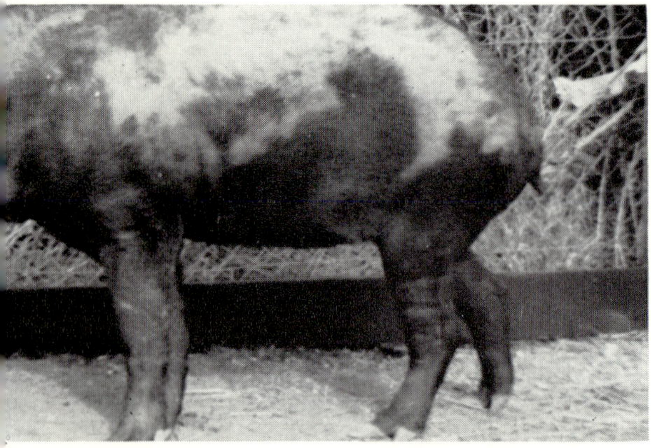

7 Few zoos manage to keep delicate Langurs alive, but at Twycross Zoo Park they have had young

8a and b. Caging: ancient and modern—both impoverished environments

its continued existence. One magazine printed an article about London entitled, 'Beasts' Belsen', and the highly respected *Ecologist* magazine started its editorial with the statement: 'The London Zoo is a shameful establishment where wild animals, living in totally inappropriate conditions, are exhibited like postage stamps in an album'. Perhaps this was a little harsh, yet anyone who is acquainted with London Zoo will, in all probability, recognize certain truths in the statement. There *is* an undeniable stamp-album mentality about the place.

We have a high regard for the Zoological Society. Its annual publication, *The International Zoo Yearbook*, is the most important document in print for those involved in zoo management and the work undertaken in the animal hospital and in the laboratories is rightly respected across the world. In 1962 the Society established the Wellcome Institute of Comparative Physiology for research into animal reproduction and in 1964 the Nuffield Institute of Comparative Medicine for studies in infectious diseases, biochemistry, radiology and pathology. In these areas London is the undisputed innovator of ideas and positive research. It seems to fail lamentably when it projects itself to the public however, and this must be due to its origins. London Zoo was born in the nineteenth century and its inexperienced yet dedicated founders established a collection containing as many different species as it was possible to obtain. Having set this precedent, the Society has been virtually obliged to live up to its reputation of being the largest collection of wild creatures in Britain. As a result it has become something of a cumbersome dinosaur. Today over 500 permanent staff are needed to care for the 6,000 animals housed there and so, despite a gate of nearly 2 million visitors a year, the cost of running this giant constantly outstrips its income. Although the emphasis today is on fewer species with larger groups of the species kept, the process is slow, and London no longer seems to be the leader in the development of zoo techniques. Frankly it is difficult to foresee how it can possibly lose its conventional zoo image.

So if the general public do not provide a case for the zoo and if many zoos fail to promote the correct policies, why do we need them? In this age the zoo has become an essential part of every civilized community. Before this century it might have been possible to justify keeping animals in captivity with words like curiosity, novelty, even amusement. But today the practice can only be tolerated if supported by those three over-worked words: conservation, research and education.

Conservation, or preservation as it should be termed in relation to captive breeding, should be the zoo's main function. Breeding to ensure that wild stocks are no longer depleted is preservation at the most elementary level. Few sincere zoologists would condone further capture of wild animals for zoological exhibition except under special and controlled situations. Breeding to save a species from extinction is the highest endeavour to which a zoo can aspire. Neither objective is so simple or straightforward as it might initially appear and, unless supported by adequate research, such programmes are destined to fail. As for education, this is essential if people are going to learn to care about the future of life on earth and, for many, zoos present the only opportunity they will ever have to see and appreciate living wild animals. This shop-window side of zoos is an important part of their role in society, but exhibition must be tempered with good taste and a responsible approach towards animal welfare—not exhibition for exhibition's sake. If correctly presented to the public, the animals can then be considered ambassadors of their species; they virtually make their own plea for continued existence. The following example will, we hope, illustrate this point.

In November 1976 one of us was commissioned to take a collection of animals from various European zoos to a new zoological park in north-west Nigeria. As the group comprised all African species, including black rhinoceros, elephant and pygmy hippopotamus, the operation perhaps sounds a little bizarre. The sad fact is, however, that, although there is still wildlife in Nigeria, it is terribly depleted and rarely seen except in areas like the 800-square-mile Yankari Game Reserve. After a 300-mile journey across the savannah, the animals all arrived safely at their destination and it was here that the fun started. As the convoy of vehicles slowly made its way through the town, people began to line the streets shrieking with delight at the sight of this wonderful Noah's Ark. One poor fellow riding a Moped was so astonished when a large elephant on a lorry passed him that he drove straight into a ditch. These people were filled with a mixture of emotions at the arrival of the animals; some were frightened, others simply over-excited—all because none of them had ever seen an elephant or rhinoceros in their lives! In Europe we might find this surprising, but it is essential to realize that Africa no longer abounds with wildlife, at least not as it did at the turn of the century, and visits to game reserves are not possible for many of the poorer natives as they have neither the time nor the money to go to them. By the same token, how many Englishmen have seen a badger or stoat in the wild, or can put a name to even the commoner species of our songbirds?

The arrival of these animals at this Nigerian zoo was a significant event for several reasons. It is now possible for the local people to visit the zoo for a very small admission charge and view the animals that were once part of their ancestors' every-day existence. The educational facilities enable them to learn about the animals, and gradually, it is hoped, they will develop a respect for their indigenous fauna by increased exposure to it. This is vitally important in a country like Nigeria, for despite the many laws of wildlife protection, poaching is rife and the market-places still abound with animal skins, crocodile handbags and snakeskin belts. It is to be hoped that the natives, having once seen a living leopard or serval cat, will eventually react to the sale of the skins, but at present these items bear little relation to creatures they have never seen. The remnant fauna clearly cannot hold out much longer in the prevailing situation, so it is imperative that attitudes change quickly.

So much for the shop-window function of the zoo. Behind the scenes, this zoo's policy is to establish breeding groups of African fauna (already they have a thriving breeding group of the rare pygmy hippopotamus) and eventually reintroduce many of the specimens into the surrounding countryside. To this end a wildlife reserve of approximately 450 square miles has been established alongside the zoo and several species of antelope, like eland and sitatunga, have been settled there. It now seems likely that as a result of this intelligent piece of preservation work the long, silent hills and valleys in one small area of Nigeria may once again ring with the multitude of voices of its returned wildlife.

But this is preservation of African animals in Africa. Can British zoos make any contribution to wildlife preservation with species other than their own natural European species? They can, always providing that the zoo philosophy we keep talking about is behind the venture; and in so doing, they can unearth much that may eventually benefit not only wildlife, but man also. The breeding of gorillas provides such an example.

The gorilla is the largest of the great apes, and there are two sub-species found only in tropical Africa—the lowland gorilla (*Gorilla gorilla gorilla*), from the lowlands in the west of Equatorial Africa, and the mountain gorilla (*Gorilla gorilla beringei*) from the lowlands and mountains of Central Africa. The mountain gorilla is the rarer, with an estimated world population of about 5,000; the lowland gorilla population has been put as high as 40,000 (although 10,000 might be more accurate). These figures are somewhat speculative but, if taken as a rough guide, one can see that the creatures' future looks depressingly bleak. The main threat to their survival is habitat destruction, and annually they are driven further into small isolated pockets of forest, constantly threatened by the work of foresters and farmers who follow in their wake. Zoos, too, have taken their toll. The manner with which the diminishing population of gorillas has been squandered for marginal returns can be truly appreciated with a single hypothesis on gorilla capture.

Out of, let us say, 48 attempted captures, perhaps 30 would survive; the local dealer would pay about £10 each for these, and would re-sell to a European dealer for the sum of about £200. Of these 30 gorillas, only about 12 might be expected to survive transportation to Europe, so the dealer revalues the animals at £2,000 in order to cover losses and freight costs. Finally the apes are purchased by zoos and, of the 12 that arrived in Europe, 6 survive to maturity. Total dead: 42; and, if none of the surviving 6 breed, extinction is escalated by 100 per cent. But, remember, the break-even point when no depletion of stock occurs requires the surviving 6 to have 48 progeny; 42 to replace those lost initially and 6 to replace themselves. In such a slow-breeding species this is totally impossible. Admittedly this is a hypothetical example of what was taking place several years ago, but it is not exaggerated. In many cases the mortality rate was 100 per cent, and the story with the other great apes follows a similar vein. So although the breeding of these rare creatures is much to any zoo's credit, do not be overawed by the contribution to preservation, for the zoos of the world are under obligation to ensure the survival of many such species, if only because of their past greed and stupidity.

In the early days gorillas fared badly in captivity. The first two specimens to reach Europe arrived towards the end of the last century, one living for 13 and the other for 17 days. Carl Hagenbeck said they died of 'home-sickness'. Gorillas certainly did not seem to do well in zoos at the beginning of this century, but gradually, with improvements in diets and accommodation, longevity records began to improve. Yet no babies were produced.

On 22nd December 1956 the world's first zoo gorilla was born and raised successfully in Columbus Zoo, Ohio, and on 23rd September 1959 the first gorilla to be born in Europe, the famous Goma, arrived in Basle, Switzerland. There are now reckoned to be over 400 gorillas in captivity, yet in the period between 1956–1975 no more than 136 births were recorded, with only 105 infants surviving. This is nowhere near good enough, yet it does show that there have been considerable improvements in knowledge and in the management of the species. Infant mortality is still in the region of 25 per cent, but noted authorities like George Schaller put the infant mortality rate in the wild at about 50 per cent. What these figures fail to reveal is that some zoos are stretched to their maximum potential, whereas others have made little effort to propagate the species. Hence the modest results to date.

The largest single collection of gorillas in Europe belongs to Mr John Aspinall and is kept at his estate, Howletts, Bekesbourne, in Kent, where there are at present 17 animals. Largely as a result of this man's devotion to his animals, a breeding group has been established and in 1976 no less than 5 gorillas were born. Four of these survived, one was reared by its mother, the remaining three were hand-reared. Realizing the disadvantages of hand-rearing, Aspinall was quick to reintroduce these infants to a group of younger gorillas, where it is to be hoped they will grow up with all the benefits of a normal gorilla social upbringing, rather than a totally perverse humanoid indoctrination. At the time of writing, other births are expected with the Howletts' gorillas, and by the time this book is in print, it is certain that there will be several more young gorillas residing in the county of Kent.

Bristol Zoo's gorillas have also reproduced and, most significantly, the baby was reared by the mother. Again this is encouraging, because hand-rearing has to date been the rule rather than the exception. The numerous disadvantages of hand-rearing will be dealt with in more detail in Chapter 6.

On 15th July 1976 London Zoo's female Lomie gave birth to the Society's first gorilla, but the father was Bristol's male, Samson, and not London's famous Guy. Guy's failure to mate with Lomie was no cause for surprise amongst many of us in zoo circles. For 22 years this magnificent male was shut away in solitary confinement, never seeing another gorilla, let alone one of the opposite sex, so the introduction of a wife in 1969 was nothing more than an act of sheer optimism.

Jersey Zoo, headquarters of the Jersey Wildlife Preservation Trust, is, so far as we are concerned, the most valuable zoological centre in Britain and its islands. We make this rather dramatic evaluation on the grounds of its youth. The zoo was started in 1959 and the Trust was formed in 1963, and both suffer from limited financial resources; yet this organization has shown more initiative than any in the British Isles. Jersey has two female gorillas, Nandi and N'Pongo, and a male called Jambo. The male is of special interest because he is the brother of Goma (the first European birth) and the result of a breeding agreement between Basel and Jersey. Unlike his elder sister, he was reared by Achilla, his mother. When Jambo arrived in Jersey he was a proven fertile animal, for he had already sired a young male by his sister, so the zoo's management had good reason to be optimistic about the outcome of his relationship with their two females. But zoo biologists, if wise, temper their considered opinions with evaluations that many dyed-in-the-wool scientists might consider disgustingly anthropomorphic. For instance, the old saying, 'marriages are made in Heaven' is a cautionary statement well worth heeding when dealing with wild animals. To believe that all one needs to do in order to breed any animal is put a male and female together is as arrogant as it is naïve. Man is not alone in his likes and dislikes: many species contain individuals with specific 'turn-ons' and 'turn-offs' with regard to relationships, be they sexual or otherwise. Jersey found this to be unnervingly true.

Gerald Durrell gives a fine description of the female reaction to their husband in his interesting book *The Stationary Ark*, but a brief account of the situation is as follows. Nandi, who was expected to be quite bloodyminded when confronted with a male, turned out to be the infatuated virgin. One glimpse of the young Adonis was enough to make her fall in love with him. N'Pongo, always considered the intellectual, the stoic, the one who could be relied upon to act in a reasoned and sensible manner, found Jambo absolutely repugnant. Jersey has therefore had to live with the incompatibility of its gorilla group. But N'Pongo has proved receptive to Jambo when she is in season and Nandi is simply besotted with him, so, to date, November 1977, seven gorillas have been born. This is a superb achievement by any standards, but, on the negative side, the infants have all been hand-reared. To explain why this is so we must resort to our own opinions. These may be wrong, but we feel that they are a reasonable evaluation of a specific situation. The clear existence of aggression between the male Jambo and the female N'Pongo results in an unnatural degree of tension within the complex. The grouping of two females and one male is abnormal. A normal gorilla group would comprise a lead male, several subordinates and a number of breeding females and offspring. The cage, although very well-considered, is too small, and as we have already indicated, it contains too few representatives of the species concerned.

Of course it is relatively easy for us arrogantly to sit in judgement, it is quite another thing to offer a constructive solution to the problem. Ideally one would like to see a gorilla enclosure many times larger than the existing complex, containing a larger and more natural grouping of the species; such would seem the perfect social environment for young gorillas to be raised in. But to achieve this Jersey would need two things: several hundred thousand pounds for the construction and several healthy gorillas that had been legitimately obtained; in other words, specimens that were not poached from the wild. In a really enlightened zoo world, the latter would not be so difficult to obtain; a number of English zoos that have gorillas are clearly not responsible enough to be allowed to own these endangered creatures and it would therefore be pleasing to see them confiscated and placed in the capable hands of organizations like Jersey.

As for finance, the same maxim seems to apply. A society that really cared about the future of the world's wildlife would ensure that its money, in the form of entrance fee or annual subscription, would go to organizations like the Jersey Wildlife Preservation Trust because they are actively involved in the preservation of certain species of wild creatures. Yet our society is not so discriminating—it pours its money into any locally convenient wildlife park or zoo—and thus genuine preservation groups are left to sit with their begging-bowls.

But to return to the relative value of Jersey's gorilla breeding. Unlike many zoos, this one is not content to wallow in the glory of an unusual breeding result; it knows that there is far more to a preservation programme than babies. Hand-rearing is, undeniably, not ideal, but one cannot leave a creature as rare, as valuable and—at the risk of sounding sentimental—as wonderful as a baby gorilla, to die. So if obliged to use human foster parents, one does everything possible to learn all one can about being a gorilla. In this area Durrell's group have inspired work of more value than any scientific programme on anthropoids has hitherto considered. The Jersey infants have been constantly watched over by veterinary surgeons and doctors, with their food intake and weight gain carefully analysed. Perhaps more significantly, in terms of comparative psychology, Jersey has encouraged studies by students from the Department of Psychology, University College, London, in the manipulative and social skills of gorillas compared with human infants. By studying such factors as the evolution of play and social behaviour, and relating them to the adult social behaviour of the species, some astonishing things may be discovered. For example, the plight of the captive gorilla (bearing in mind that captive breeding might eventually prove to be its only life-line) is beginning to throw some light on the child-battering syndrome that is such a problem in our modern society. Solve one and one might solve both. The intimate relationship between zoological and medical research is irrefutable. Already it seems likely that recent research work undertaken by London's Wellcome Institute in monitoring the reproductive physiology of the great apes will not only help to ensure the survival of these endangered animals but may in turn throw further light on human reproductive processes. This important research programme, which at present is basically a system of hormonal monitoring that enables a relatively accurate pregnancy diagnosis, constitutes a vital part of successful wildlife management and is the result of close co-operation between the Jersey Wildlife Preservation Trust, Bristol Zoo and London.

With the present gorilla population, it should be possible for zoos to maintain a consistently high population of these apes, and with co-operation and organization between zoos there should be little difficulty in producing a sound breeding programme that will ensure the survival of the species. Unfortunately, until now gorilla breeding projects have been the result of private agreements between the more reputable establishments—for instance, Basel/Jersey and Lincoln Park, USA/Howletts—but these arrangements are too few and undisciplined. Further advances seem unlikely unless more zoos operate in conjunction with others in an open, frank manner and are prepared to divulge details of animals in their possession. It is surprising how secretive many zoos are. Some would apparently have us believe that they never have a sick or injured animal, and that none of their stock ever dies! This stubborn wall of silence hinders progress considerably.

Jeremy Mallinson, Zoological Director of Jersey, attempted a census of the great apes in British zoos during the summer of 1976. He wrote to each zoo asking for details of the species kept and the age and sex of each specimen, with a view to establishing a central fund of information that could be used to further breeding developments in the UK. However, one of the commercial groups had bluntly refused to co-operate, nor would they give any explanation for their rejection of his modest enquiries. One must assume that they feared their animals being placed on record, as these statistics might eventually reveal the true mortality rates and accurately determine their consumption of wildlife.

The zoo is a sadly misunderstood institution, but there is no reason why this should be so, for surely its purpose, its objectives, are very clear. The Nigerian zoo that is endeavouring to return the native species to their natural home, to educate the local people in wildlife appreciation, to study and breed the endangered animals of West Africa, illustrates what zoos are about. The breeding of gorillas at Howletts and Jersey—the important research work on gorilla reproduction and psychological development which may one day provide the basis of the first programme to reintroduce gorillas into the wild—is what zoos are about. Their common duty is to study wildlife in order that the knowledge so gained can be put to practical use in wildlife preservation and education. And they should be looked on as custodians of living creatures, not as possessors, for the zoo is under an absolute obligation to its inhabitants. Every zoo employee, from the director down to the youngest keeper, should remember this: they work for the animals' benefit first and foremost; all other considerations are secondary. Anything that falls short of these ideals is not a proper zoological establishment and is therefore unacceptable.

The suggestion that exhibiting wild animals simply for profit is a legitimate practice provided the animals are well cared for, is quite ludicrous and morally wrong. If a species is to be adequately cared for in captivity, it is necessary to analyse its physical and mental needs, dietary requirements and its normal behaviour patterns, then develop a captive environment that is biologically appropriate. To achieve this one needs experienced staff and a substantial amount of capital to construct the eventual enclosure and finance the necessary research. With all that, the profitability of the scheme dissolves into a fine mist. Knowing this to be so, it is clear that every year the majority of British zoos indulge in a contrived deception of the visiting public. They fill their guidebooks with shameless lies about their non-existent research work and their educational programmes. With every birth they rush to the nearest newspaper office with a story of the supposed contribution to wildlife conservation. Most of it is mere fantasy.

To create a zoo, animals have to be captured and transported from their country of origin: they can easily die in the process; many do. Animals have to adapt to a new environment and if this is ill-conceived they can suffer permanent physical or mental damage, which many do. If they reproduce in captivity, it is vital that the offspring are healthy in body and mind. This will depend, largely, on their social and physical environment and other factors, like diet. But because of inadequate research into their specific needs, a large number of zoo-born animals are not normal.

It must therefore be obvious that it is in the ultimate interest of the zoo as an institution that the capture, maintenance and breeding of wild animals are handled solely by those who have the technical and practical experience necessary and, equally important, a high-minded regard for the well-being of each individual creature. The question is: will the British zoo-keepers and those to whom they are responsible sort out their problems and determine their objectives while the animals are still around?

Chapter 3

# Capture, transportation and quarantine

For each wild animal one sees in a zoo or wildlife park one can almost guarantee that over the years dozens of the same species have suffered the rigours of capture and transportation from their place of origin. This is the most terrifying and traumatic experience of their whole lives. Capture and transportation from the wild for supply to zoological centres is of course quite different from capture in the zoo for transportation to another wildlife or zoological park. There is a distinction between the methods required to achieve either successfully.

### *Wild capture*

Wild capture can be accomplished by a variety of techniques and although each specific technique has clear benefits with one particular species, invariably it will be found to be absolutely useless with another. Some of these methods the reader will probably find abhorrent, but one must begin by understanding that there is no such thing as an entirely kind or gentle method of capture. The child grabbing a slow-worm or lizard for its home menagerie does not intentionally hurt or frighten the creature; but the effect of being seized and unable to escape most likely causes the unfortunate animal intense terror that can result in death. This analogy holds good for any capture of any species under almost all circumstances! So, once again, we are back to our hackneyed insistence that the end justifies the means. If the purpose is of consequence to the species concerned, then the method can be justified, but if the end is merely to line someone's pocket, the action is, without doubt, barbaric and cruel.

Of all methods of capture the pit is certainly the most ancient, and a modification of it is still used today. A pit is dug across a track known to be used by a certain animal or herd of animals.

The bottom is covered with a net, on top of which is placed a thick layer of moss, grass and any other vegetation that will break the animal's fall. When the animal falls into the trap a previously prepared crate is pulled from the undergrowth and the animal hauled into it. Depending on the species concerned, the technique is not so cruel as it might at first appear. A giraffe so caught would almost certainly break its legs and probably its neck as well. A cat—for example a lion, tiger or leopard—would more than likely be very annoyed and little else. So, when using the method, it is vital that one knows precisely what is going to be caught.

Netting is a more modern concept, but it has its weaknesses. Basically the netting procedure is similar to that of the pit. One finds a pathway regularly used by the species and spreads a folding net across it. Failing this, the net is placed in an area into which it is possible to drive the animals. Our own experience on netting is largely confined to zoo specimens, but as a general rule we have found a tendency for the nets seriously to bruise or cut flesh. Therefore, unless the operation can be accomplished with reasonable expediency, it is usually a rather rough and unreasonable practice.

Mist nets are generally accepted as a standard method for the capture of small birds. As the name implies, they are extremely fine, rather like gigantic hair nets, that are spread across an area known to be traversed by the smaller avians either during morning or evening flights, or migrations. The birds become entangled in them by their claws or feathers, and are quite helpless until released by the human operator; they are rarely able to injure themselves. In fact the system is almost foolproof, providing the nets are not left unattended for any length of time. If they are, the birds will have a very slow, long-suffering death, brought about by a mixture of fright, hunger and thirst.

With small mammals, the wire trap is still popular because it is so remarkably effective. This trap is not so vicious as it sounds; it is not designed to clamp a paw in a vice-like grip; it does not hoist an unwitting creature off the forest floor to suspend it by its neck until the arrival of the trapper. Quite simply, it is a wire tunnel with two movable ends that snap shut once an animal is inside. The trigger mechanism is a wire platform upon which is placed the bait and once the animal steps on this it activates the trap, both ends snap shut and the animal is captured. To use this device effectively one must be thoroughly acquainted with the behaviour of the species and the routes they use. Failure to visit the trapping site at regular intervals will naturally result in the death of the captured specimen.

Burrowing species, once located, are usually caught by the method of blocking the exit hole and then simply digging them out. Dangerous reptiles like poisonous snakes are generally located by herpetological experts, who quickly seize them with a long stick on the end of which is a padded noose. They then secure them in a canvas bag. These creatures rarely suffer undue stress or fear, because they have a rudimentary brain structure and because they are usually caught by highly experienced individuals. Perhaps the price of fear has its advantages.

Running down and roping is a technique that has become more streamlined with the passing of time. Nowadays it is still considered an acceptable method of capture, but in reality it has improved little. In the early days running down was achieved on horseback. The desired specimen was chased until it became so weary that it was possible to sling a rope about it, then a second and possibly third line was attached and the animal was held fast until a crate could be brought. Today the process is speeded up and jeeps or landrovers are used to chase the animals. The resulting furore can do much to clarify the choice of specimens. If any specimen is infirm or suffering from an injury or some illness, it can be expected to collapse under the pressure of pursuit, and is not therefore suitable for capture. A healthy specimen can be pursued until exhausted, then roped and crated. Of course, it can still collapse and die as a result of the stress caused by this rude method of capture, although sometimes the damage done might not reveal itself until later in the animal's life.

The growing knowledge of how to immobilize, anaesthetize

or tranquillize wild animals has brought an entirely new dimension to wild animal capture and there is little doubt that the use of chemical compounds provides a quicker, safer and more humane technique than conventional methods; but only with certain species.

The conventional capture of a rhinoceros would go somewhat as follows. Three or four vehicles would set off in pursuit of the rhino and endeavour to hem the animal in whilst at the run. This wild cross-country race would go on for as long as it took to get a lasso about its neck with the help of a long pole. Once roped, the rhinoceros would go frantic and charge the lorries again and again until further ropes could be got round its feet and it was thrown. Once the feet had been securely tied, the rhinoceros would be winched up on to the truck and, like a trussed chicken, bounced across country to the base camp where it would be released into a secure compound.

As one might expect, such a method can have profound physiological effects. The chase causes intense fear and the subsequent exertion leads to a state of physical and mental exhaustion. There is usually a dramatic rise in heart rate and respiratory rate and a rise in body temperature that, with many animals, is accompanied by profuse sweating. The increase in the respiratory rate can cause a washing-out of carbon-dioxide, producing a condition known as hypocapnoea, which in turn affects blood circulation. There is a raising of blood pressure which normally speeds up circulation. However, if the chase has inspired extreme fear, the breakdown products in the muscles accumulate faster than they can be removed and capillary stagnation begins to take place. In some cases this leads to a condition known as white muscle disease. This can be likened to a form of cramp where the blood supply to an area of muscle or muscles is very much reduced. Because the condition is so extreme the muscle becomes permanently damaged. It is in fact an early symptom of the state of shock and indeed is a common cause of death in newly caught wild animals, especially in the antelope species. (This condition is quite different in origin from the white muscle disease sometimes found in British calves; the latter is due to vitamin E and Selemium deficiency, and is of dietary origin.)

The alternative to this harsh form of capture is the chemical technique. Using this method, the catcher, a qualified vet, would estimate the weight of the animal and load a projectile syringe with a suitable quantity of a powerful narcotic. This syringe is fired from a specially manufactured rifle and it discharges the drug on contact. The drug most generally used is a narcotic up to 10,000 times more powerful than morphine, which takes effect within a couple of minutes; an antidote is used to revive the animal later. The rhinoceros would be approached cautiously and darted from a safe distance, the only sensation experienced by the animal being a sharp prick, followed by drowsiness and then unconsciousness. Totally immobilized, the rhino would be carefully examined by the vet prior to being loaded into a crate and injected with the antidote. Within something like thirty minutes one rather bewildered, but uninjured, rhinoceros finds itself captive.

This does make it sound as though drug capture is the only method of capture worth considering, but of course there are drawbacks. Different drugs will have different effects on many species of animal and even closely related species will show distinct variation in tolerance. There are also side-effects and after-effects with immobilization. For instance, body temperature, respiration and blood pressure can be affected, so it is imperative that the animal remains under the supervision of experienced veterinary staff for some time following the administration of the antidote. Great care must be taken at each stage of the operation.

Other disadvantages of immobilization include problems such as the losing of a darted animal. This is usually the result of a poorly handled attempt at capture. Too small a dose administered in an area with plenty of cover can allow the darted animal time to flee some distance before collapsing. If not found quickly it will almost certainly die or be attacked by the first predator which stumbles upon it. On the other hand, some species are too small or delicate to be darted whilst others, because of the denseness of the habitat in which they live, could not be hit with a dart in a thousand years. Predominantly arboreal species if darted will naturally flee to the highest branches, from where they eventually fall to their deaths when the drug takes effect; whereas a creature like the hippopotamus has been known to dive below the surface of the water when hit and drown as it loses consciousness.

So the chemical technique has fantastic advantages over other methods with certain animal types, but with other animals its use is quite impracticable. And, in addition, although it should go without saying that all wild animal capture should only be attempted by experienced people, chemical restraint is a very specialized field which should only be entered by experts, for the exact dosage is vital.

A novel method of capture that does not involve the use of ropes or drugs has been devised in Africa in recent years. Whole herds of animals are gently driven between walls of nothing more substantial than coloured plastic which lead into corrals or waiting lorries. This exercise in applied animal psychology makes it possible to catch several hundred animals at one time without injury or undue stress, but as far as we know it has so far only been used by game departments for the re-location of wildlife into suitably protected areas.

Lastly, there is the most disgusting and wasteful method of all —killing the parents in order to gather the infants safely. In the last chapter we mentioned the capture of gorillas and the high mortality rate resulting from transportation to the zoo. Yet the killing began at the point of capture. Gorillas are devoted parents who will protect their infants to the death. As the capture of adults was an almost impossible task, the hunters shot them and removed the babies from their still warm bodies. Orang-utans too fell victim to this practice. Two factors influenced the method of capture. Orang-utans are only found in dense, almost inaccessible forest which is usually some distance from human habitation of any size. The outward trek, therefore, was generally quite difficult but the return journey with the captured animals was a doubly arduous and slow affair. Secondly, orangs are both agile and elusive yet, if cornered, even a half-grown ape will prove to be strong, pugnacious and extremely dangerous. So, as an adult can weigh between 75–100 kg., it was decided that they were too bulky to transport over the rough terrain, besides which their capture was far too risky. Thus it became standard practice to slaughter whole families in order to obtain the more handleable infants. These pitiful little red bundles of fur were stuffed into crates and taken to one of the many coastal villages to be found in Bornea and Sumatra, thence to Singapore and onward to their final destination in Europe or America. Hundreds of these babies were dead long before reaching the native village, but many more died en route to the western world and the full death toll ran into thousands. It has been estimated that for every live orang-utan to reach a zoo or private collection, at least six had sacrificed their lives, yet even this horrifying calculation is probably well below the true figure. These practices are outlawed today and the scientific zoos and the International Association for Zoo Directors have declared an absolute ban on the importation of orang-utans, save those few specimens which have the export permission from the land of origin. These are usually infants confiscated from poachers or unauthorized persons found to be keeping them. Yet there is little doubt that illegal shooting continues and the illicit trade in young orangs is still going on. It is not a flourishing trade, merely a trickle by comparison with twenty years ago, encouraged by the less scrupulous animal fanciers who are unconcerned about purchasing smuggled animals. As late as 1966 over 200 babies were offered for sale on the world market, but today, although the figure must be smaller, it is difficult to produce an accurate census as these sales are effected far more discreetly in a kind of underworld of wildlife transactions.

The anxiety caused by any of these methods of capture is, of course, quite apparent, the reasons less so. But the act of capture is only part of the operation; we will call it Phase One.

Phase Two is the movement of the freshly caught animal from

the point of capture to the base camp, and there is a dilemma here. Should the animals be rested for an hour or so before taking them to the camp-site, or should they be loaded on to the truck immediately, in whatever state they appear to be, and moved as fast as possible to base where they can be given shade, water, food and veterinary treatment if necessary? Experience has shown that the latter is preferable, but obviously certain precautions should be taken.

In tropical climates capture should not be attempted in the heat of the day and the hot dry season should be avoided at all costs. In so far as it is possible, the actual capture should take place reasonably close to the base site so as to save the specimen unnecessary exhaustion from a long journey over rough ground shortly after the trauma of capture. It is also commonsense to avoid capture of animals that are rutting or in oestrus. Knowledgeable catchers will take all this into consideration and also be able to distinguish between those that have been chased too hard or are reacting too violently to capture and make an accurate decision as to whether or not it is advisable to move the animals to base camp or release them immediately in order to ensure their lives. A really well-organized catching team will in fact include a veterinary surgeon well acquainted with the psychological disturbances associated with capture as, if considered necessary, he can administer drugs to reduce shock and aid circulation. Also at this time, in responsible circles, it is considered normal routine to inject a broad-spectrum antibiotic, because the stress of capture and transportation predisposes the animal to infection—indeed, pneumonia and enteritis are common occurrences at this stage in the operation.

Phase Three is the settling period at base. Here the animal is released into a compound, or at least a larger cage, where it can be kept quiet, and be rested. Theoretically, the animal slowly recovers from its frantic ordeal, yet it is only met by new ones—acclimatization to capture and the natural fear of being tended by men. In this area no amount of scientific theory can replace the indefinable rapport that some people have with animals.

Love, understanding and patience are the qualities needed to give any wild creature confidence. Admittedly they should be tempered with a degree of biological and ethological knowledge, but no amount of learning can produce them. As the successful gardener is undeniably credited with 'green fingers' without the benefit of scientific sanction, so the true animal attendant can achieve a unique relationship with his charges, despite the hostile condemnation of academic predictions.

In the first fortnight the animals will usually lose weight and refuse to eat. There are exceptions of course. We know of several animals that should have died of terror when caught, yet when they arrived at camp they simply proceeded to gorge themselves. One particularly engaging marmoset of our acquaintance (marmosets are one of the smallest and most delicate of all primates), having been captured by a South American native, sold in a market place and eventually transported to the United Kingdom—a procedure that must have encompassed a period of several months—finally emerged from its box in a state that can only be described as high indignation as opposed to fear. Neither afraid nor impressed by the presence of human beings, he started to screech from the moment of his release and refused to stop until a bowl of fruit, far in excess of his needs, was placed before him. Having sated his appetite, he looked at us with an attitude of benign tolerance, and in time, when his cage had become his own special territory, he severely bullied anyone who dared to trespass within his preserve. Animals are terribly unpredictable.

To return to generalizations, in time the captured specimen will gain confidence, begin to eat, and put on weight. The period usually allowed is 30 days, but in our opinion this should be doubled at the very least before moving the animal further. Once it has recovered from the capture, is putting on weight, and the veterinary officer is satisfied that it has undergone a sufficient period of post-quarantine, then arrangements can be made for the final journey to its eventual zoo home. The method of transportation from the country of origin is identical to the transportation between zoos, apart from the fact that the newly-caught animal is more likely than the zoo specimen to have to adjust to an alien, inhospitable, climate. So we will leave crating and shipping for the moment and deal with this aspect of the business in conjunction with transportation between established collections.

We have outlined a few of the main methods of catching animals, emphasizing that the procedure inevitably involved fear and physical stress for the individual involved. To our minds this irrefutable fact conjures up one question alone. Why catch them in the first place?

Controlled wildlife management produces a valid reason for the act. The wardens and rangers of Africa's national parks are required to perform a variety of functions, one of which is the study of the ecology of each area of the park and the effects of the indigenous fauna on the available vegetation. Conditions in the protected area may require the movement of certain animals from one area to another or even one park to another, if a population has built up beyond the carrying capacity of the range. Zebras have frequently been moved from an over-grazed area to one of greater abundance in the South African parks, for example, but only in order to keep the environment and its inhabitants healthy. The process must cause a great amount of stress, yet no one who truly cares about the future of wildlife could realistically object to the operation. The alternative is habitat destruction and slow starvation for the animals.

Beyond the relocation of wildlife to areas of greater security and food abundance by conservationists, nearly all animal capture is concerned with providing stock for captivity and this cannot be condemned out of hand—each case should be judged upon its own merits. To adopt the sloppy animal-lover's creed that all wild things should be left alone to live their lives in the peace and security of their homelands, would be to place the seal of doom on many species of wild animal. Man has left little of the natural world unmolested, he has bent the delicate chain of life quite mercilessly and caused considerable damage to both the environment and its inhabitants. Hope of redressing these crimes rests largely on advised tampering with the works in order to return things to a level of normality. In other words, if a wild animal sits on the crevasse of extinction and has been placed there by the activities of our species, it would be irresponsible to leave it to fall—any legitimate effort to save it is better than nothing. Protective legislation is a step in the right direction but this alone will not necessarily ensure the survival of the species. In the first place, laws are easily broken. To assume that merely to pronounce an animal protected will lead to an end of its persecution would be optimism indeed. Africa, Asia, Australasia, Europe, Eurasia and the Americas all have animals that are officially protected by law, but all discover cases of the illegal taking of these species, dead or alive, every year.

The African elephant is a good example of what we mean. The latest surveys by wildlife experts have revealed that the elephants of Uganda and Kenya could face extinction within the next generation, primarily because of the value of their ivory. In the 1960s Kenya's ivory exports were running at about 90,000 lbs a year but by 1975 the figure soared to an estimated 900,000 lbs, worth about 14 million dollars annually to the Kenyan economy. It is estimated that in Kenya alone poachers are killing between 15,000 and 20,000 elephants each year; as the total Kenyan elephant population is reckoned at about 120,000, it does not take a mathematical genius to realize that at the present rate of extermination all the elephants will be gone by the early 1980s. Ironically, Kenya's famous Tsavo National Park has suffered terribly from a very singular problem—too many elephants. As early as 1954 Tsavo East's Warden, David Sheldrick, drew attention to the alarming increase in Tsavo's elephant population and the subsequent destruction of the woodland. Somewhat predictably, no one listened. In 1963 an ecological survey on the park concluded: 'If the habitat is to be preserved in its present form . . . the numbers of elephants will have to be controlled'.

Three years later the Ford Foundation financed a three-year project under the directorship of Richard Laws. Laws' proposals included the recommendation for the slaughter of a further 2,700 elephants, but the park trustees overruled him and things were left to take their course. Over the years Tsavo changed, vast tracts of thick woods are now reduced to sparse grassland, and most of the larger trees eaten by the elephants are dying out. In 1971 the tragedy anticipated by Sheldrick 17 years earlier happened. A severe drought hit Kenya and the Tsavo elephants, having destroyed most of their food resources, began dying in their thousands. Thus, today, when the future of Kenya's elephants rests on controlled park management, what was the finest elephant sanctuary in the country can support less than 25 per cent of the animals it could have held comfortably 20 years ago, and these few are being ruthlessly poached.

Africa's endemic disease, poaching, is also reducing Uganda's magnificent parks into vast elephant's graveyards. Kabelega Falls Park, that had an estimated elephant population of over 14,000 in 1973, reckoned that the surviving herds totalled only 2,448 specimens in the winter of 1976, and this was a slight improvement on the 1975 figure of 2,246. Rwenzori National Park had nearly 3,000 elephants in 1973, but in late 1976 the actual figure had been reduced to 704 specimens.

The poachers use a variety of methods to kill the elephants, from poisoned darts to bananas laced with battery acid, methods that are effective in that they eventually kill the animal, although the elephant may suffer a painful and lingering death that can last for as long as three weeks. The Ugandan authorities are genuinely concerned but have been powerless to stop the killing. In 1975 President Idi Amin brought in a law that all elephant poachers would be shot, yet this harsh decree has not yet been enforced, possibly due to the difficulty in apprehending the culprits. Worse, it is widely reported that the President's own troops, using high-powered weapons, are killing elephants and their calves on a grand scale, thereby setting a deplorable example to the rest of the population.

We are not suggesting that in this instance captive breeding is the answer to the elephant's dilemma. In fact we have grave reservations about the suitability of these creatures for captive maintenance. But clearly legal protection alone is not enough. It must be backed with careful wildlife management and stronger penalties for poaching that are made effective by intensive security measures. Alternatively, or let us say in addition, with some species it would obviously be advantageous to establish a nucleus of captive animals for the purposes of research and breeding, rather than trust their future entirely to the dictates of successive governments or the caprices of legislative protection.

The very best catching expeditions are unquestionably those promoted by zoos or wildlife trusts with a view to obtaining specimens of one or two specific species (not dozens) that are appropriate to the collection's objectives. Preparations for the trip would start several months before the actual departure date; permission is obtained from the country of origin to catch and export the animals, the staff study all available literature they can on the species to be obtained, a base camp, transport and shipping dates are agreed upon, and so on. By the time the actual expedition is under way, plans for almost any eventuality have been laid. The most significant feature of such expeditions, however, is not simply the careful preparation, but the fact that the senior members of the team involved in the capture are also the people who will accompany the animals on the journey home and care for them for possibly the rest of their lives. In the eyes of the expedition, these animals are not merchandise but representatives of a species they wish to study in order to establish reliable data that can be put to constructive use in zoology, ethology, biology and, if things go exceedingly well, can be used to ensure that new management techniques are available to aid the species' survival. Parents are not killed, animals are not chased until they collapse from exhaustion, every care is taken because the initial motivation is engendered by a concern for wildlife.

Expeditions of this type are not just flights of fancy, they do occasionally take place. Gerald Durrell's collecting trips are typical of this kind of operation. Readers of Durrell's books will be acquainted with the many adventures he had during his travels, yet his amusing anecdotes on the problems of animal capture actually serve to obscure the more serious side of his work. In his youth Durrell might have been termed simply a collector whose trips provided a source of income and a vast fund of material for his books, yet even 20 years ago the qualification would have been ungenerous. Time will show him to be one of the most important conservationists Britain has ever produced and the work of his Wildlife Preservation Trust will be discussed in a later chapter. The later expeditions of Durrell, carried out in the interest not just of fresh material for a book, but more specifically on behalf of the Trust, have proved invaluable in terms of research and preservation in captivity.

To take one example, in December 1964 the BBC Natural History Unit approached Durrell with the idea of making six half-hour programmes about an animal collecting expedition. Never slow to take advantage of a situation that could benefit the Trust and his beloved animals, Durrell accepted: the 1965 Sierra Leone expedition was launched. The objective: to obtain specimens of the threatened pygmy hippopotamus, colobus monkey and rare West African leopards, in addition to some of the commoner but less often kept species like palm civets, pouched rats and brush-tailed porcupines. The expedition failed to acquire pygmy hippos because they found it too protracted and complicated a job for the time available, but the remainder of animals on their list were secured.

There is no doubt that the exercise was justified, for wildlife protection in West Africa is wanting, to say the least; in 1965 it was desperate. Gerald Durrell explained the situation in the *Second Annual Report* of the Jersey Wildlife Preservation Trust and his words describe not only the problems of conservation, but also the frustration he felt at being thwarted in his objectives. He states:

> The conservation picture in Sierra Leone is somewhat depressing: the Government is fully aware of the necessity for National Parks and for the protection of the indigenous fauna,

but they lack both trained individuals to do the job and the necessary money. This leads to a rather anomalous situation where nearly every animal found is strictly protected on paper but is not protected in fact. A good example of this is the Banded Duiker, Cephalophus zebra; I wanted to procure a male for Frankfurt Zoo as they have three females and it is essential that such a rare Antelope is established in captivity. My application to capture a male was turned down on the score that it was the most strictly protected animal in the whole of Sierra Leone. I would have felt that this was most heartening, if I had not, at every village up-country, been offered Banded Duiker skins for sale.

However, the ursine black and white colobus brought back by the expedition have become a most valuable breeding group, and their behaviour, dietary and medical requirements have been carefully documented since the moment of arrival.

That responsibility tempered with the enthusiasm of the catching team is not only undeniable, but should perhaps be emphasized, and the colobus produce the most positive evidence of this responsible attitude. Durrell managed to procure seven black and white colobus (from which the breeding group has been established) and eight red and black colobus. The latter are not only rare, but so uncommon in captivity that they would command an incredibly high price in the animal market. Not that in this instance they would have ever been offered for sale. Durrell felt that there was not sufficient time to get them used to captivity before moving them to Freetown, Sierra Leone's capital, to be sure of their survival, so the monkeys were released. Other trips by Durrell's Trust follow a similar pattern of moral rectitude.

Expeditions of this kind are certainly worthwhile but it has to be emphasized that they are very much in the minority. As a general rule, animal capture and shipping is the preserve of the wildlife exploiters rather than the preservationists, for most zoo specimens are obtained from the animal dealer (see Chapter 4).

*Transport between zoos*

Earlier we said that the transportation of animals from the wild to a zoo was essentially similar to transportation between one collection and another. Differences do exist: the newly captured animal, having just started to get accustomed to the strangeness of captivity, has then to undergo a second ordeal—the rigours of re-crating, travel and the problems of acclimatization in an unfamiliar climate. As it is also only just beginning to get used to the presence of men, it is unlikely to feel particularly comforted by the close proximity of a number of them crowding about its crate, something they are obliged to do when manhandling or securing it. Even when offered food and water the animal is liable to panic simply because its natural impulses tell it to flee, but restricted by the walls of its crate, it is unable to do so.

We know of many cases where this has happened, but perhaps the clearest illustration of the problem is an incident that occurred in the early 'seventies when several impala antelopes were loaded on to a ship in South Africa for transportation to Europe. Apparently the animals had settled quietly in their crates in the hold of the ship when a young, inexperienced native attendant was sent below to water them. He accidently dropped a bucket, causing a sudden, violent noise, and two antelopes broke their necks in their panic.

Captive animals, particularly if they have been well cared for by attentive keepers, can cope with the situation with remarkable equanimity. The writer well remembers being charged with the responsibility of moving a pair of young adult Siberian tigers a distance of some 150 miles, and the worries that accompanied the task. Both animals were captive bred, and during the six months they had been in his care proved themselves not just friendly, but loving. Nevertheless, their reactions to being boxed and trundled across Britain were, at best, open to conjecture. Crates were accordingly tailor-made—a mammoth one for the huge male and one of more modest proportions for his mate. Then they were placed in position in front of the cage slides and the operation began. Contrary to everyone's expectations, both cats walked unconcernedly into the crates and remained quiet throughout the journey, even to the extent of making the typical tiger noises of contentment and rubbing against the front of the crates whenever the writer approached them. When introduced to their new quarters they defied all the rules of feline behaviour and romped about like a couple of youngsters on a seaside outing. No attention was paid to the new environment and no reaction could be ascertained as to the effects of transportation, for in honesty the reaction was nil.

We have to confess at this point that we are cheating a little because we have taken two extremes; highly nervous impala that had, in all probability, endured a traumatic capture and period of settlement in captive conditions, and two tigers so accustomed to the peculiarities of men that little they did caused them concern. But the obvious distinction emerges. As a general rule the newly captured specimen can be expected to be far more nervous and delicate than the old lag who has become familiar with the experience of travel. However, apart from these points the principles of transportation remain the same.

Crates are not only essential, they are generally preferable to loose loading, for animals that are loaded in groups can be expected to excite or even injure one another as a result of the unfamiliar experiences of travel. But a crate is only any good if built for the individual concerned, and if the crating that precedes transportation is accomplished by humane means rather than barbarity. If we resort to yet another example, the reader will possibly appreciate the point we are trying to make.

Fairly recently, we were required to ensure that several animals were transported safely from Europe to Africa, that they arrived alive and in good health. Three zebras were amongst the consignment, one from a zoo in Holland, two from a collection in England. The specimen from the Dutch zoo arrived at the rendezvous point in a crate that was a perfect example of what good crating should be: large enough to allow the zebra reasonable space, whilst at the same time, restricting movement sufficiently to prevent it turning around, thereby trapping and injuring itself, and strong enough to resist its kicking and breaking the container. At both ends were slides that could be opened fully, or, by the simple removal of a catch, half opened so as to facilitate access to the animal in an emergency, without the risk of its escaping. The interior was lined with padding to avoid grazing and the floor slatted to prevent the animal slipping. Feeding and watering could easily be dealt with as a manger with covering slide was attached to the front of the crate. As for the zebra, it was as calm as could be because it had been fed into the container for several days prior to transportation; therefore, when the slide was eventually shut it experienced little more than mild surprise.

What a difference in the zebras from England! Two crates arrived; one so large that the inmate could move almost two feet back and forth, and could turn its head to such an extent that its becoming wedged in a half-turned position was more than a distinct possibility. The other crate was so small that the zebra's rump was forced against the rear slide whilst the head and neck were cramped at a horrifying angle by the front slide. Neither crate was padded and there was no way of feeding or watering the animals. Even worse, the larger container that allowed so much freedom of movement had rudimentary slides made of plywood without any type of reinforcement. One good kick would have knocked it out. The cramped container was rotten, not only in the obvious sense: it was literally green with mould. This animal had been driven to Harwich, shipped to the Hook of Holland, driven on to Amsterdam, where it was eventually to be loaded on to an aircraft belonging to an IATA member. During its subsequent journey this zebra was seen by several officials, including a Dutch Government veterinary officer, but not one of them saw fit to remark on the deplorable container or protest at the obvious discomfort of the animal inside it.

In the meantime, the problem of watering was surmounted using a watering-can; both animals quickly learned that a bright orange plastic spout appearing between the slats of the crate meant water, and feeding was achieved by cutting food and hay into small pieces and pushing it through the gaps in the slide. Although successful, these methods were extreme and should not have been necessary in the first instance.

The crating operation that preceded their departure is of even more interest. On the date of departure the zebras were supposed to be loaded on to the vehicle taking them to the docks at 3 p.m. This would mean that the crates were ready for loading far earlier than this. At 2 p.m. the animals were still standing in their stalls, dosed with a tranquillizer called ACP, notoriously ineffective with zebras. Shortly before the departure time the man in charge of catching allowed the first zebra into the race leading to the crate and with much shouting and beating with sticks and iron bars, he and his assistants secured the zebra in its container. Five minutes later the panic-stricken creature had to be released again as it had managed to turn a cart-wheel in the crate and was lying on its back, quite helpless. The second zebra was boxed with the same delicacy as the first, then the first zebra was re-crated and the shipment set off. Whether these people truly expected any animal to travel well after such an experience or whether they just could not give a damn is really immaterial, so obviously are they misplaced in their responsibilities.

Capture and transportation, whether from the wild to captivity or between collections of captive animals, is a process that must be tempered with kindness, compassion, understanding and consideration of the problems that might be encountered en route. Without wishing to labour the point, two black rhinoceroses on the same trip were contained in exemplary crates into which they had been quietly fed for over a month; their favourite indulgence whilst travelling was to be scratched on the head. They came from the highly respected West Berlin Zoo. Two elephants in the same shipment were difficult to control and, despite constant attention, remained anxious throughout the journey; one actually smashed the top off its crate. Crated in obviously inferior containers within hours of transportation and never having a chance to settle down, they remained problematic throughout the journey. Their source, almost predictably, was a safari park in Italy.

It is often suggested that the kindest method of transporting animals is to resort to tranquillizers, but frankly we reject this except in extreme circumstances. To appreciate what using a tranquillizer means we have to digress into an element of scientific fact. The more technical name for tranquillizers is neuroplegics, from the Greek meaning a blow or paralysis of the nervous system. The name describes the reaction; they produce central nervous depression, which is simply a dulling of consciousness and a reduction of spontaneous motor activity. The extent of this depression is determined by the sinking of the eyes, drooping of the eyelids and in most animals a protrusion of the third (protective) eyelid. Blood pressure and the metabolic rate fall, and there is a tendency to become sleepy. Large doses will affect the animal's motor system and they will tend to show muscular weakness, swaying of the hindquarters and there is a likelihood of the animal lying down or even collapsing. Doubtless they feel rather like an executive on an international flight who has over-indulged himself in rich food and alcohol, pretty sick and disorientated. Highly nervous animals will occasionally have a sort of paradoxical or inverse reaction to the drug and may become excitable, uneasy and increasingly difficult to control. Canids are especially susceptible to this kind of behavour; wild or domestic dogs may snarl and whine continually, even bite themselves severely. Also they take a variable time to react; usually 45 minutes is sufficient for these drugs to take effect, but occasionally the reaction may be delayed for several hours, and they last for one to three hours, though not infrequently much longer than this. A dangerous factor of tranquillizers in relation to transportation is that they often affect the animal's heat regulating system. This means that if it is placed in a cold environment there is likely to be an uncontrollable drop in body temperature, conversely, if held in a warm area the animal's temperature will rise to an alarming level, causing sweating, fainting and dehydration. Perhaps we should also emphasize at this point, without getting too technical, that although the drugs in the tranquillizing group are closely related chemically, they do have a wide variety of different actions. Some have distinct depressive qualities whereas others are particularly useful in suppressing disturbances like nausea. All can react quite differently on different species of animal.

Tranquillizers are just not the panacea to the problems of transportation that so many imagine them to be. They are of undeniable value in cases of extreme stress or to combat the effects of shock, but they should never be used on an animal known to be suffering from anaemia, renal damage or circulatory disturbances. Thus they should only be administered in the presence of a qualified veterinary surgeon who can carefully monitor respiration and blood pressure and who, if either falls too sharply, can apply treatment.

Of course, this is rarely the case, and the tranquillizer has gradually become a means of covering up the incompetence of the persons in charge of crating. If the crate is too large, too small, too insecure, or the animal has been violently over-excited during the crating operation, 'Give it a tranquillizer' is the cry that goes out. The disadvantages are self-evident—if any animal collapses in the container during transportation, the odds are against getting it up again. Nevertheless, it has become common practice for every Tom, Dick and Harry to administer these drugs even without veterinary advice. How many creatures die as a result is a point for speculation. However we do know of several instances where large mammals have collapsed and died in their crates and in the absence of alternative evidence we are inclined to believe that they died as a result of indiscriminate drug administration.

In our view the standards of catching, crating and transporting are in sore need of upgrading, attention to containers should be improved, the indiscriminate use of drugs clamped down on, and animals should always be accompanied by experienced keepers when travelling. In this, as in every aspect of maintaining wildlife in captivity, there is no substitute for loving care.

*Quarantine*

The importation of wildlife into the United Kingdom is not allowed to take place indiscriminately. Restrictions do exist and some of these are quite severe. Quarantine is the restriction with which the reader will be most familiar, especially in relation to the horror at present threatening our shores—rabies. All species of animal—wild or domestic—that are potential rabies carriers are required by law to serve a six-month period in an approved quarantine area upon entering this country. The cage must be rodent and bird-proof, double-wired to prevent accidents and it should also have double security doors that make escape impossible. Actual physical contact between humans and the isolated animals is totally prohibited because rabies is not just transmissible by a bite; an infected animal's saliva on an area of broken skin, as might happen when a dog licks its attendant's fingers for instance, can prove sufficient to pass on the infection.

Hoofed animals also have to serve a quarantine period, although in this instance there seems to be a certain illogicality about the present regulations. Exotic ungulates arriving from distant countries could easily bring with them a whole host of diseases and viruses that could be transmitted to our domestic livestock. Thus the statutory twelve months' quarantine (part of it behind fly-proof screens) is undeniably a prudent measure. Twelve months is long enough to allow for the incubation of most infections and should one develop during this time the precautions taken, like the burning of all bedding within the quarantine area and disinfectant trays for staff to walk through when entering or leaving the premises, would ensure that it remained contained in that one area. But in our view these regulations take on a somewhat ludicrous form in relation to imports from Europe. A giraffe or antelope coming from Europe into the UK still has to be quarantined for the twelve-month period, yet a domestic cow has only to undergo a 28-day period of surveillance. This seems an absurdity. We know of a case when an Asiatic ox that was born in a European zoo was brought into the UK within hours of a shipment of cattle—the cattle came from a farm less than ten miles away from its birthplace—yet they had only to serve the set 28-day period whilst the 'wild' ox had to be quarantined for a year. What makes this seem so stupid is that of the two species the cattle, having been kept on open farmland, were the most likely to have been exposed to infection and, when released on the English countryside, the most likely to spread it. Zoo specimens, on the other hand, are kept in permanent isolation from domestic animals in the confines of the zoo.

The main argument against tightening the regulations on domestic hoofed stock seems to be one of finance. If farmers had to quarantine freshly imported blood-stock for twelve months (presuming suitable quarantines were available in sufficient numbers), they would be faced with alarming fees in addition to the initial purchase price of the cattle. British farming just could not take this and presumably the imports of fresh breeding-stock would dry up completely. But despite the apparent inconsistencies in the regulations there are few responsible zoologists who would contest the necessity of quarantine, for we are extremely fortunate in this island, in that we are relatively free of diseases that can become epidemic in a short space of time. In the past quarter of a century the complacent British have occasionally been shaken by the odd outbreak of foot-and-mouth disease; in the early 'seventies pandemonium nearly broke out when rabies was suspected in Surrey. Yet by comparison with many other countries these incidents are so trivial that we might consider ourselves in Utopia. Therefore, let us abide by the regulations and safeguard our complacency.

However, having cast our vote in favour of accepting the rules designed to protect man and wildlife, we are bound to say that the idiosyncrasies of legislation are cause for grave concern. Bob Dylan the folksinger/poet put the legislative system into a nutshell with the line: 'Money doesn't talk, it screams'. And in our opinion this pessimistic view does justice to our Government's attitude towards any form of control. In recent years the trade in wildlife has mushroomed into a multi-million pound industry which makes it not a leader, admittedly, but an important contributor to the financial stakes of the United Kingdom, and in this context we accuse the British Government of turning a blind eye to what it sees as an irrelevant problem of sentiment, yet a relevant source of income.

However, to return to practicalities. If one genuinely wishes to establish a species in captivity it is of primary importance to settle the animal in its new environment as soon as possible. The perfectly justifiable rules of importation deny this. A newly arrived animal is obliged to suffer a period of time in an environment that is not only unfamiliar, but generally ill-conceived in terms of the animal's requirements. Once animals arrive in the importing country, survivors should at least be able to look forward to proper care and treatment, but the restrictions of quarantine frequently mean that a creature that has suffered the agony of capture, settlement and transportation is then forced to bear the restrictions of quarantine accommodation. For some specimens, this can sometimes prove the final nail in the coffin.

We have always found quarantines most unpleasant places, not because they are badly run, but simply because movement is so restricted. In the early 'seventies we had reason to be associated with one that had been established in Salford, Manchester, and as quarantines go it was a pretty impressive affair. The building was a newly built warehouse in the heart of an industrial estate, suitably isolated from any area containing domestic livestock and large enough to hold a considerable number of animals without overcrowding them. The warehouse was a long rectangular building with an entrance hall and offices at the front, whilst at the rear a large sliding door and retaining yard permitted the loading and unloading of stock of any size without undue risk to the handlers or animals. Spacious pens were constructed along the length of the walls, and three larger ones were built in the centre of the building, leaving enough space for walk-ways for the staff and the movement of crates. Each pen was provided with a slide communicating with the next pen, so that animals could be moved to allow for cleaning without causing them undue agitation, and the floors of each pen were lined with non-slip tiles for cleanliness and safety. When in operation (sadly it closed down after only one shipment), the quarantine housed giraffes, scimitar horned oryx and roan antelopes to name but a few. It was spotlessly clean, the animals were well cared for, staff were in attendance 24 hours a day, and the Ministry's regulations were observed to the letter. It was an exemplary set-up that showed how quarantine restrictions can be adhered to without causing the animals an unreasonable degree of deprivation. Despite its obvious quality, it was nevertheless unpleasant to watch its magnificent inhabitants pacing about in their few square yards of concrete isolation, condemned to remain there for 12 whole months without access to fresh air or sunlight.

So the price of getting some of these animals into Britain is exceedingly high. In certain cases a period of 18 months or more may have passed between the moment of capture and the actual introduction of the animal to its permanent zoo environment. And if this is in any way sub-standard, one must conclude that the exercise is costly in terms of animal suffering and negative in objective. The animals involved in the quarantine we have mentioned were more fortunate than most. Nearly all of them came into England from other zoos so they had not recently undergone the procedure of wild capture, and they were all imported on behalf of specific zoological establishments and were headed for a definite location. In this case the majority of animals went to John Aspinall's marvellous collection at Howletts in Kent and the Marwell Zoo in Hampshire, where the enthusiast will find one of the largest groups of scimitar horned oryx in captivity, bred largely from this initial importation. But many wild animals imported into Britain for the zoo trade are destined for the dealer's yard, there to languish until someone meets the price asked for them!

No existing quarantine in the United Kingdom seems to match the qualities of the now deceased Salford establishment, and since few importations seem to be as specific as this one, most imports must be unjustified.

Chapter 4

# The wildlife trade

Over the years the dealer has obtained a reputation somewhat akin to the slave-trader; and it is a reputation many rightly deserve, for like the slavers they travel widely, enjoying the not inconsiderable profits of their venture, whilst their cargoes of horror lie concealed below decks, hidden from sight so as to disturb few consciences. Exceptions occur, but we reckon that one could count the worthwhile dealers on the fingers of one hand; to the rest a dead animal is only a financial loss and they give scant thought to the suffering it endured whilst dying. A few years ago whilst in Europe one of us had occasion to visit a dealer, whose reputation was supposed to be good. After dinner he suddenly announced that he had a consignment of animals in transit at the airport and invited an inspection. Naturally it was accepted. On arrival at the airport he was told by one of the staff that the animals had been placed in a small room at the end of a cargo building which was heated and draughtproof, but that some of them did not look very well. The dealer thanked him and without apparent concern set off in the direction indicated. The animals turned out to be Celebes black apes (*Cynopithecus niger*), an uncommon species closely related to the macaques and found only on the Island of Celebes and a few adjoining islands. Ten of them were packed tightly into two crates roughly the size of tea-chests, and it could be seen immediately that three were already dead. Putting on a pair of thick leather gloves the dealer opened the flaps at the base of each box and started to pull the corpses out when suddenly he must have noticed the expressions of horror. 'Don't look so concerned,' he said, 'with the price I have got on this deal I shall still make a good profit.' It is with regret we have to say that, at that time, the authorities felt that as the animals were in transit and had neither been crated nor technically received by the dealer, prosecution was out of the question. Today this delightful gentleman is still in business and often sells stock to British collections. From all accounts he is doing very well financially.

Dealers also seem to show little regard for how the animals are obtained or whether the capture is legal or not. Where Durrell failed to obtain a permit for the capture of one male banded duiker, a really sharp dealer would have succeeded irrespective of whether this would have been a good or bad thing to do. The method is simply good, old-fashioned bribery.

Any reader who has travelled in some of the developing nations, especially on the continents of Africa or South America, will most likely have had the uncomfortable experience of being baulked by officialdom. A small discrepancy on a visa or health certificate, for example, can pose great problems for the unwary traveller and he can easily find himself in a whole heap of trouble. But it is usually possible to get out of this dilemma, for if you are quick enough to pay the right official, and his assistants of course, an agreed sum of money, they will happily overlook the irregularity and allow you to proceed. In some countries this bribe system is almost accepted practice, so much so that one of our colleagues recently suggested that if he travelled with enough money in his wallet there were several countries he could enter without a passport! Knowing the situation, the animal trader frequently turns it to his advantage. By approaching the right ministers and officials he simply buys his export permission and another endangered animal slips into the commercial pipeline.

One dealer with whom we are acquainted went so far as to suggest that, given a contract at the right price—presumably high—he could obtain almost any animal we desired. During the conversation he started to talk about obtaining giant sable antelope (*Hippotragus niger variani*) and said that he had great hopes of such a capture in a forthcoming expedition he was planning. At the time there were approximately 160 sable antelope (*Hippotragus niger*) in captivity and some were breeding, but no zoo in the world has specimens of the giant sable and therefore their breeding potential in captivity is unknown. We emphasized this point but he reacted predictably by saying: 'I am aware of that fact, so you can imagine what each will be worth'. Desperate to avert his reckless campaign, we outlined the basic facts about the species. The giant sable actually stands rather shorter at the shoulder than the typical race but is normally distinguishable by the immensity of its horns. The total population may be between 500 and 700. The precise distribution of the herds is difficult to ascertain, but the majority of animals are found in the southern part of the Luando Reserve in the Malange district of Angola, and they are completely protected by the Government of Angola, with severe penalties for infringement of the law.

None of this seemed to perturb him and he intimated with many a nod and a wink that if certain people were looked after, he would get his animals. He had even thought of the problems of documentation, pointing out that if he captured young animals with little horn development no one would be able to distinguish between them and the typical race. Thus he could legitimately move them into Europe as sable antelope and by the time the deception was discovered it would be too late for anyone to do anything about it. But perhaps his final remark was the most revealing. He said: 'Anyway, if I'm rumbled I can always plead ignorance and insist that I thought I was getting ordinary sable'.

We doubt that our threat of exposure if he carried out the operation either deterred or frightened him, but thankfully this catching trip that would have involved poaching, smuggling, false documentation and bribery never took place. However, the meeting endorsed our feelings about dealers, for his remarks were in earnest and there is no doubt that similar things are happening across the world every year.

To appreciate why we hold most dealers and traders in such contempt, one has to understand that our opinions are not based on incidents like the one we have just mentioned alone. We have visited many of the places where dealers keep the animals after arrival in the United Kingdom, and that is generally a never-to-be-forgotten experience. However, transportation, like the act of capture, is a procedure that causes fear and stress; it must be dealt with by responsible people who care about the individual animal if it is not to turn into a nightmare of suffering and death. And that is precisely what it has become, largely as a result of the dealers' activities. The size of the trade alone is quite incredible, and if we look at just one port of entry into Britain the reader will glean some idea of the horror involved.

Shortly after the Second World War the international trade in animals began to increase, and by 1952 over 100,000 creatures were passing through Heathrow Airport each year. The increase was directly attributable to the speed of air travel, which was far preferable to the long sea voyages previously endured, and the growing demand for wild animals as pets, subjects for scientific experiments and zoos. This demand has not decreased, nor has it begun to level out and today a little over a million animals pass through the airport every year. The RSPCA felt, and indeed still does, that most animal passengers are making a journey that is not strictly necessary, in fact that three-quarters need not have travelled at all. And, considering that the majority of arrivals are there merely to gratify someone's private whim, or are destined for laboratory research, they condemned the trade without reservation. The animals still came by the crateful of course, and it was because of the distressing state of the creatures arriving at Heathrow that the RSPCA built a hostel there in 1952 so that the animals could be rested, receive treatment if required and, in many cases, be re-crated before continuing their journey. It was a pioneering venture well ahead of its time, and is still in use today with very little alteration. Basically it consists of a long service hall into which vehicles can be driven to facilitate the unloading of crates. There is a kitchen, a food storage room containing a great variety of feedstuffs to provide for the incredible diversity of creatures that pass through its doors, and quarantine kennels for the 7,000 or so dogs accommodated annually. There is a room where fish containers (they are usually packed in polythene bags) can be reoxygenated and reptiles can be cleaned out and watered. There are two large multi-purpose rooms, one with two strong loose boxes for holding various creatures from horses to young elephants. One room has a removable floor, beneath which is a heated tank where seals and sea lions can be released, another has spacious cages for monkeys, a third is designed to house tropical birds. Near the offices there is a small surgery where sick or injured animals can be examined and treated, and adjacent to this is a sick room containing smaller cages where they can be kept warm and under observation. Until recently it was the only hostel of its kind in the world and, despite recent developments, it is still the best. Kennedy Airport in the USA have a small facility for housing animals and Schipol Airport in Amsterdam has a small hospital with heated rooms where animals are temporarily stored in their crates. But these facilities do not compare with the British complex.

Unfortunately, since the opening of the hostel at Heathrow, it has been the unpleasant duty of the staff, year after year, to remove thousands of dead bodies in the course of cleaning out travelling boxes. Many of the arrivals are found to have been neither fed nor watered since leaving their country of origin, and all too often the crate proves to be too small, wrongly constructed or inadequately ventilated. The problems encountered at this hostel are really too numerous to list; nevertheless, the fact emerges that even today few consignments arrive here that do not contravene existing regulations.

Smuggling accounts for some of the most obscene cases found. For instance, one box that arrived at Heathrow caused the hostel staff some concern as it had only a few airholes, thus making it impossible to see inside, and although the accompanying documentation declared that it contained a cat, the noises issuing from it were anything but feline. On opening the container they discovered six dead and dying monkeys. On another occasion a consignment of falcons was discovered concealed in the bottom of a crate, their eyelids sewn together so that they could not see and would therefore travel in silence. Another practice unearthed from time to time is the crating of live birds—usually pigeons—along with birds of prey, to act as food on the journey. It is an absurdly stupid and cruel act because predatory birds rarely feed well when travelling and are certainly unlikely to make a kill whilst crated, so their intended victims suffer a needless journey in a state of constant fear.

Badly crated animals, particularly those that are unattended on the journey, all too often die. To take just one case, a young lion sent from Amsterdam to Karachi, via London, was found dead at Heathrow; the metal container was smaller than the overall size of the animal and the ventilation was inadequate. The veterinary post-mortem showed clearly that the lion had died from suffocation and heart failure: just one of the innumerable victims of the trade in wildlife.

Admittedly we are moving into an area that is not solely concerned with zoo animals; the thousands of finches, parrots and tortoises that have died in transportation over the years can be largely blamed on the pet trade, the many monkeys that died in the 'fifties were mostly on their way to laboratories. But zoo animals have died, and will continue to die, because of their neglect by both dealers and zoos.

Regulations do exist. In the early 'seventies the International Air Transport Association, whose main function until then had been to regulate passenger transportation and cut out unfair competition, decided to meet animal welfare organizations like the RSPCA, and discuss standards for the transportation of animals. As a result, the IATA *Live Animals Regulations Manual* was drafted but, because it was only an advisory document, only the more responsible airlines took notice of it. The manual has been revised on a yearly basis and the 1976 edition (the 5th) has been made mandatory.

Unfortunately this really means very little, and the document has had only a slight effect on the standards of air transportation, for the Association is self-policing and imposes only trivial fines on its members. Evidence presented to the IATA Board must be presented by their own inspector who has to see the case for himself, and so far as we are aware, no case has yet been considered by IATA. Even if a case were pursued, the fine would be unlikely to injure the carrier financially. Alternatively, the airline could simply withdraw or resign from the organization to avoid punishment. The few laws that do exist on animal transport vary from nation to nation and are applicable to all transporters of animals, whether they are IATA members or not.

Nevertheless, the manual is a distinct step forward. It contains good general observations on transporting animals, including guide-lines on health and hygiene, climatic effects, feeding and watering, crating, loading, sedation, documentation, the importance of clear labelling, even a small section on animal behaviour —points that those responsible for transporting animals should already know, but apparently do not. As knowledge improves, so the manual can be revised; yet in its present form it does seem to have one major drawback. Most of the animals involved in the wildlife trade are being transported from developing to developed nations, and quite apart from the fact that the manual is printed only in English, its general format must appear terribly complicated and involved to those people constructing the crates in the country of origin, usually local natives. The many ramshackle containers still arriving in Europe and the UK seem to prove this to be true, so it is clear that legislation that is applicable to the airline and its agents should also be applied to the consignor or consignee.

It is heartening to be able to say that in 1973 British legislation largely provided for this with the introduction of the Transit of Animals (General) Order. As its name implies, it is a general order, and uses terminology such as 'inadequate', 'unnecessary suffering' and phrases like 'feeding and watering must be adequate and at suitable intervals', without specifying what they are. Intentionally ambiguous, this order requires a specialist opinion to ensure enforcement in each case, and although this method has its weaknesses, it does give the animal the benefit of the doubt where doubt exists. Briefly, this order places a general duty on carriers and other persons in charge of the animals during carriage to ensure that the provisions are applied in a manner appropriate to the species. A further duty is imposed on those concerned with loading and unloading to ensure that these operations are carried out in a manner that will not cause the animals injury or unnecessary suffering. Such persons are also required to guard against the risk of injury or suffering arising from a number of other specified causes. Other provisions deal with feeding, watering and the general care of animals during transit, and with their accommodation in ships, aircraft and on vehicles. The loading or carriage of pregnant or unfit animals is prohibited except in rare circumstances, when permission has been granted by a Veterinary Inspector of the Ministry of Agriculture, Fisheries and Food. It also provides for the way in which animals injured in transportation are to be dealt with.

Recently, under this legislation, a number of prosecutions have been taken; most have been against airlines and most have been successful. Although the airlines have grumbled, there is every reason to believe that much more care will be taken in accepting animals for transportation, unless properly contained, and it should mean that the airlines will in turn put pressure on the agent in the country of origin to be more attentive to the conditions of shipment. Already several airlines have shown themselves unwilling to accept 'job lots' of unaccompanied animals and a particularly interesting refusal to load was brought to our attention. An elephant sent from a British collection for transportation overseas was turned away from an aircraft at Gatwick by the plane's captain on the grounds that the crate was unsafe and the elephant in too great a state of distress to undertake the strain of several hours' flying. Sadly the poor elephant had to suffer the long drive back to the zoo, but perhaps in future the park management will ensure that their animals are properly crated and settled down before transportation.

Quarantine has an unprecedented side-effect: inflation of prices. Anyone who has brought a pet into the United Kingdom and had to place it in a quarantine station will be acutely aware of how expensive this can be; quite apart from the heartache of separation the person importing the animal is faced with a pretty startling monthly bill. Magnify this several hundred times and one has some idea of the astronomical costs of quarantining some of the larger ungulates.

Consider a giraffe in a British collection. How can one arrive at a price for such an animal? In the country of origin there is the initial cost of capture, maintenance for several weeks at a base camp, and shipping fees to the United Kingdom. By the time it arrives in Britain a considerable sum of money stands in a crate, facing a further year of investment without any financial returns from its exhibition. Quarantine fees mount up rapidly, and staff salaries, food and veterinary expenses have to be accounted for. Should the animal die, all this capital is irretrievably lost. Thus a giraffe settled in an English collection might be valued at about £2,000 to £2,500.

One would naturally assume that even if the sole motivation behind the importation was one of profit, the financial commitment involved would prompt concern about the animal's welfare. After all, a dead animal is a financial loss. In part this is true. The largest importer of giraffe into the United Kingdom in the last decade is the Chipperfield Organization (the Chipperfields are the biggest animal dealers in the world, with an estimated annual turnover of more than £500,000*) and we have no evidence to suggest that this group has ever wittingly neglected any of its giraffe imports. However, it has shown little hesitation about allowing its giraffe to cross-breed!

To explain our criticism—there are eight sub-species of giraffe:

| | | | | |
|---|---|---|---|---|
| 1 | Nubian giraffe | *Giraffa* | *camelopardalis* | *camelopardalis* |
| 2 | Kordofan giraffe | *Giraffa* | ,, | *antiquorum* |
| 3 | Chad giraffe | *Giraffa* | ,, | *peralta* |
| 4 | Reticulated giraffe | *Giraffa* | ,, | *reticulata* |
| 5 | Uganda giraffe | *Giraffa* | ,, | *rothschildi* |
| 6 | Masai giraffe | *Giraffa* | ,, | *tippelskirchi* |
| 7 | Angola giraffe | *Giraffa* | ,, | *angolensis* |
| 8 | Cape giraffe | *Giraffa* | ,, | *giraffa* |

*J. Chipperfield, *My Wild Life*.

For many years scientists described these sub-species as distinct species, a theory that has now been abandoned because all of them can cross-breed and have fertile offspring. It is possible to cross-breed a male lion with a female tiger, or vice-versa (presuming one has a suitably warped mind), but the offspring will prove infertile because the parents belong to two distinct species; but sub-species will breed and produce young that are fertile. Nevertheless, the existence of a sub-species, or regional variation of a species, is the product of time and environmental pressure. The human equivalent is the Negro of Africa, the South American Indian, the South Sea Islander, the Eskimo—all belong to the species *homo sapiens*, all will cross-breed and produce fertile youngsters, yet all have developed independently in their own unique geographical isolation. Physical appearance, social customs, language, are all distinctive to the race, and the same is true of wild animals.

When one applies these facts to wildlife it is clear that preservation of the species rests almost entirely on preservation of the sub-species, and one might reasonably ask if cross-breeding is compatible with conservation. Indiscriminate cross-breeding is leading us towards a situation where the child's elementary reading-book that states 'A tiger' or 'A giraffe' might actually become a reality, and we could be left with a host of creatures that bear only a rudimentary resemblance to the animals that once existed. But are these points of consequence to the safari parks? When kept in herds, giraffes are easily bred and a giraffe (type unspecified) is worth possibly £2,000; and the same applies to many other species.

A second method of covering the cost of importations is to obtain large numbers of a species at a ridiculously low price to cover the mortalities involved in transportation and settlement. These can then be resold in Britain at a substantial profit. This is rarely possible with the larger or rarer species, but it is common practice with birds and reptiles, and again it is a point on which we shall elaborate later in this chapter.

Beyond the quarantine regulations there are other restrictions that affect the importation of wildlife, but these are of a different nature, and are intended to protect the endangered species from exploitation rather than the British from infection and disease. In a world where mankind is rapidly destroying the environment and its wild inhabitants, any legal measure designed to cut down the slaughter must be welcome. The Achilles heel to the concept is whether paper legislation alone is enough. Unless rigidly enforced it may be worse than useless. To a certain extent this seems to be true, and as far as we are concerned the flood-gates of wildlife importation are opened and closed, not as benefits the species concerned, but as suits the whims and pockets of the operators. This accusation is unashamedly strong, it hints at bribery and corruption, or at least to the convenient turning of a blind eye by the authorities when enforcement of the law poses greater problems than bending it. We have in fact stumbled upon incidents that prove this to be so. One of us was shown a letter from a Member of Parliament which assured the recipient that proposed measures to inhibit wildlife exploitation would not go through. It finished with the line: 'Don't worry we will ensure that the proposed Bill is a dead duck'. We also know of a member of an animal welfare organization who enjoyed an all-expenses-paid holiday in Africa. The expenses were covered by an animal dealer and somewhat significantly the holiday-maker has never since raised objections to the operations of this particular person.

The fountainhead of all conservation activity is the International Union for the Conservation of Nature and Natural Resources (IUCN) which has its headquarters in Morges, Switzerland. It is a sort of 'doomwatch' organization which continually gathers information from all parts of the world on environmental problems: the status of rare fauna and flora, levels of pollution, and so on. Its main publication, the *Red Data Book*, is the conservationist's bible. There is one volume for each group: mammals, birds, reptiles, amphibians, fishes, trees and so on, and being a ring-file book it can be kept up-to-date by adding fresh pages as they are printed. This means that anyone who wishes to know the present status of an endangered species can have all the latest available information at his finger-tips. Valuable as this document is, it is only a service provided by the IUCN. More vital is their relentless pressure on governments to take positive steps to protect the environment and its inhabitants.

In the early 'sixties the Seventh General Assembly of the IUCN called upon all governments to restrict the importation of rare animals in order to preserve such animals from the danger of extinction, and in response to this the British Government passed the Animals (Restriction of Importation) Act in 1964. Under this Act an Animals Advisory Committee was set up and Clause 3 (2) of the Act sets out its duties as follows:

> It shall be the duty of the said Committee to advise the Board of Trade, the Secretary of State or the Minister for Home Affairs for Northern Ireland upon any question which they or he may refer to the Committee in connection with the administration of this Act.

Although an advisory committee, their recommendations on the issue or refusal of a licence are normally accepted, and these recommendations are based on a wide range of criteria. Depending on the species for which a licence application is received, the factors for consideration can vary widely, but one or two general factors are taken into account in almost all cases. These are: the status of the species in the wild: whether the animal for which the licence is required was born in captivity or captured in the wild; and, if the latter, the circumstances of its capture. Other criteria are the purposes for which the animal is required—breeding, research, exhibition, resale or pet. For those animals that are highly endangered the Committee will normally only consider recommending import if the animal is required for breeding, and even then they will take into account the importer's past record in breeding the species, or a related species, the accommodation available and details of the proposed breeding programme.

Two lists outline the species covered by the Act. List 1 details species for which licences will not normally be granted, and applications are referred to the Advisory Committee for individual considerations. List 2 details species which are licensed without restriction. Both lists are liable to amendment without notice.

This all sounds highly reasoned and well thought-out, but how effective it has been is a matter of opinion. Take the great apes, for example. All are in decline in the wild state, so what has our protective legislation done for them? In 1971 only four apes were imported and these were chimpanzees. In 1973 the Committee allowed the importation of 16 chimpanzees, 4 gorillas from the Cameroons, 14 siamangs and 8 Lar gibbons. All of these came into the country for breeding and exhibition purposes save one, which was licensed under the dubious heading of 'Pets, *bona-fide* pets of importers', whatever that may mean! It is difficult to condone these imports, in particular the importation of four gorillas caught in the wild. We do not know the method of capture so cannot criticize the actual operation, but taking gorillas from the wild is a totally unjustifiable act and has been considered as such since the mid-'sixties by nearly all conservationists. There is simply no case for bolstering existing captive stocks with further raids on the wild.

Possibly the figure of 42 great apes imported into the United Kingdom in one year does not seem much, but these animals are considered endangered and given List 1 status by the very Committee that granted the licences. When one moves on to look at the List 2 imports—that is, those animals that can be imported without restriction—the figures become truly impressive. The 1973 figures for primates for instance: *Cercopithecidae* (the long-tailed monkeys of the Old World)—32 specimens were brought into Britain for zoos, 33 for resale by dealers and 9,895 for research. *Callithricidae* (from the New World)—none for zoos, 34 for resale; 1,006 for research. The research figures are always astronomical and although we are concerned only with zoos in this book, it is impossible to side-step the subject of the three major consumers of wildlife: research, the pet trade and the fur trade.

There is a marked tendency for people to sanction research because it is deemed necessary, essential even for the health of mankind. They sanction it without knowledge of the facts, without any idea of the price, purely because they believe it to be in the best interests of man. There exists a gruesome chronicle of experiments that subject animals to intolerable mental stress or physical pain, the bulk of them performed without anaesthetic. In 1974 four and a half million fully conscious creatures were given electric shocks, blinded, infected with syphilis and gonorrhea, poisoned, inoculated, starved or force-fed, behind locked doors in Britain. Their only source of protection is a piece of antiquated legislation, the Cruelty to Animals Act of 1876. In 1975 the figure went up to a staggering five and a half million experiments on living animals, of which less than one-third were classified as 'medically necessary', and some of those are questionable. The rest consisted of tests to determine the safety of products like shampoos, cosmetics and detergents. Mascara is put into eyes to see if there is blistering or 'gross destruction' of the cornea, skin products are rubbed into flesh to ascertain whether or not they cause tissue damage or irritation. Is this justifiable?

The growing demands of scientific research are placing an increasingly heavy burden on the wild stocks of many animals, a burden that is becoming unbearable in the case of the primates. Probably half a million primates are captured annually solely for use in research laboratories. The yearly total in the USA alone has, on occasion, risen above the 200,000 mark. And the effects of this mindless consumption are rapidly being felt.

The commonest monkeys of India are the Rhesus Macaques (*Macaca mulatta*), their range extending through Northern India, Assam, Burma and China. They are frequently seen in zoos where they have always been popular as they are hardy, and are known to breed freely even under the most unfavourable conditions. Rhesus have also been favoured by medical researchers and are the primate most commonly found in laboratories, but in all the years of use no effort has been made to breed them in order to make these research programmes self-sufficient. The argument against has always been that it is too expensive to finance separate breeding work, so the policy has been simply to catch, catch and catch more specimens. Now the rhesus population, described as 'abundant' in the mid-'sixties, is showing a drastic decline in numbers.

In all, 17 species of primates are at the present time directly threatened with extinction as a result of the trade in laboratory animals. In its turn, this reduction in wild stocks means a shortage of monkeys on the market, and greatly increased prices when any are available. As costs spiral upward it is us, the taxpayers, who have to contribute millions of pounds towards the further murders of the countless animals used annually in government research programmes, most of which are not essential.

There are alternatives to using wild animals, a fact that makes the present trade all the more meaningless. Tissue culture (using human cells) and other test-tube methods have been used in many fruitful experiments, and these techniques are also far cheaper than the standard animal tests. Nor is there a shortage of human material. There are post-mortems, people who understand the purposes of research and willingly volunteer, people who are sick or injured who, in the course of normal treatment, can be clinically observed to the benefit of medical science and themselves. It should also be remembered that although medicine has benefited from experiments on animals, these do not always provide conclusive results. The Thalidomide disaster bears grim testimony to this fact, for the drug was thoroughly tested on laboratory animals and found to be safe. In fact many of the world's leading scientists have stated that if enough work was carried out to find alternatives, experiments on animals would not be needed at all. But despite the great question-mark that hangs over research, the annual sacrifice of domestic and wild animals is allowed to continue.

On 1st January 1976 the Animals (Restriction of Importation) Act of 1964 was suspended and new regulations were brought into force under the Import, Export and Customs Powers (Defence) Act 1939, a wartime measure which was supposed to have been repealed when the State of Emergency ended. Under this Act anything may be prohibited for import, export or transit with no reference to Parliament, and no private individuals may prosecute for illegal import. These controls were introduced by the Department of the Environment in response to pressure from conservationists, and as a gesture intended to bring Britain into line with the recommendations of the 1973 Washington Convention. This had been convened to establish world-wide rules on the import and export of endangered species, and lists similar to those of the IUCN's *Red Data Book* were prepared as working guide-lines.

The Washington Convention was signed by the British Government in March 1973 but they did not ratify the conditions to which they had agreed until 1st November 1976. Finally the Endangered Species (Import, Export) Bill received Royal Assent on 3rd February 1977.

Under the present Act, normal procedure is to specify a restricted product by reference to a heading of the Customs' Tariff. When drawing up the list of products to be controlled, the Department of the Environment omitted several items and left so many loopholes that these restrictions rapidly gained the reputation of a farce. Turtles prove a case in point. All eight species of marine turtle are endangered, and listed in the *Red Data Book* and in Appendix 1 of the Washington Convention (i.e., that they should not be in trade). Thus the importation of live turtles into the United Kingdom is banned, although it is still permissible to import certain turtle products. Turtle soup is obtainable from Fortnum & Mason at £2.50 a jar, for instance. Unworked tortoiseshell is prohibited, yet on the Department of the Environment's own list of selected imports of wildlife into Britain (January–July 1976) one finds that 709 lbs of tortoiseshell from the Hawksbill turtle (*Eretmochelys imbricata*) was licensed. Spotted cats of all species are declining at a terrible rate because of the activities of the fur trade. The same list covering only a six-month period shows that a total of 50,005 such skins were licensed. Reptile products provide even more alarming figures.

Nearly two and a half million skins were licensed. If the rate continued, one can presume a figure of 5 million by the end of the year.

The ruthless exploitation of the elephant is also allowed under these 'controls'. The Department of the Environment, when drawing up the list of products to be controlled, had the foresight to include elephant hair bracelets but omitted to include elephant-skin suitcases. Their measures also controlled the importation of unworked ivory yet placed no restriction on items like ivory chess sets (often valued at over £1,000 a set). In actual fact, in the first six months of 1976, Britain imported 31,096 kg. of ivory from Kenya (the plight of the Kenyan elephant is mentioned in Chapter 3); 11,488 kg. from Zambia, and in addition 19 tusks from Zambia, 40 from Kenya and 96 from the Sudan. Apparently we are to believe that whole tusks are considered 'worked ivory'!

The same vagaries apply to the skins of crocodilians. Under the 1976 measures it is illegal to import their skins but quite permissible to bring in items like crocodile-skin handbags. Somewhere along the line even these ridiculous rules seem to have been bent, for since the introduction of the new regulations over 5,000 caiman (South American alligator and crocodile) skins have been imported under licence. The species concerned is totally protected and its export banned in Brazil, but the skins concerned are listed as coming from 'South America'. But one of the best examples of all is the vicuna (*Vicugna vicugna*) a small, wild relative of the well-known llama, and its downfall has been its dense, short pelt, for the wool is a prized possession in the world's fashion houses. The animal is found in Peru, Chile, Bolivia and Argentina and all four countries ban its export, dead or alive. In 1970 the United Kingdom, in accordance with the wishes of the Peruvian Government, banned the importation of skins, and the World Wildlife Fund urged other nations to follow this example. Is it not surprising then to find vicuna cloth on sale in London at prices ranging from £275–£300 a metre? At present the onus of proof lies on Customs, who must show that the item in a shop has been illegally imported. Thus vicuna cloth marked 'Made in England' is openly sold in Regent Street, London, despite the fact that the wool and hair of this rare South American cameloid have not been licensed for import since 1970. Retailers insist that this cloth is old stock!

So far we have only mentioned animal products. What of the living creature? In the same period, that is the first six months of 1976, the Department of the Environment issued licences for an astonishing number of animals without any apparent reason or justification. Three hundred and ten caiman were allowed into the country, all listed in the *Red Data Book* and protected in many of the countries from which they were exported, as were a large number of lizards and snakes from Colombia—where they are completely protected. One hundred green turtles and 100 luth turtles were licensed in spite of the fact that they are listed as endangered by the IUCN. All the Indian pythons imported into Britain in that six-month period were of the sub-species *Python molurus bivittatus*, clearly listed in the *Red Data Book* as endangered, yet the Department of the Environment saw fit to allow the importation of 458 of them.

Tortoises brought into the country for the pet trade have always caused the conservationist great concern, for less than 10 per cent of them can be expected to survive their first winter in Britain. The average yearly importation has been running in excess of 200,000 individuals, so, understandably, new restrictions are always greeted with hopeful anticipation. In this instance the hopes seem ill-founded. In the brief period between January to July 1976 the Department of the Environment (presumably with the approval of the scientific advisory authorities) granted import licences for over 270,000 tortoises from Morocco and the USSR.

Bird imports follow the same pattern. Over 100 falcons, including 10 peregrines, were admitted primarily for falconry or airfield bird control. Thirty crowned pigeons were imported from Malaysia—none of the 3 species occur there, they all come from New Guinea where they are totally protected. And the fishing tackle industry was granted permission to import 2,000 head of Lady Amhurst's pheasant and 2,000 kingfishers from China, plus nearly 3 million peacock plumes for floral art and fishing tackle. All this in six months!

Halfway down the list is a mystery item. It seems that permits for the importation of 100 Malayan sun bears were issued. We say this is a mystery because we cannot ascertain who wished to import them or in fact if the quota was ever taken up. Why anyone should wish to import so many of this species is even more bewildering. Neither zoos nor safari parks are likely purchasers and these animals seem unsuited for use in medical research.

What is alarming is that so many of these animals and animal products brought into Britain should never have been given import licences. These fall into three categories: those listed on Appendix 1 of the Washington Convention, those listed in the IUCN *Red Data Book* as endangered and those which were originally poached or smuggled. The basic idea of the Washington Convention was for the industrialized nations which import most of the furs, ivory and skins to reach agreement with those countries which possess the greatest number and variety of wildlife. Quite simply, import controls in the consumer countries would be linked to export controls in the countries of origin, but, excellent as the idea is, it seems that government ministers never seem quite clear on how to put these schemes to good effect. To date, the British results certainly do not encourage much faith in the government's declared interest in the preservation of wildlife, and one is inclined to assume that most of the debates on endangered species consist largely of hot air. The government's duty is nonetheless clear. In this, the last quarter of the twentieth century, there are nearly a thousand known mammals, birds, reptiles, amphibians and fish in danger of extinction from two main pressures: the destruction of their habitat and a trade created by people who want to have them as pets, see them as zoo exhibits, wear clothes made from their skins or eat them. These latter pressures could be largely removed by a dramatic tightening of restrictions. This would obviously not be economically painless to implement, but unless these restrictions are implemented without undue delay there will not be any wildlife left to import.

The Endangered Species Bill was first introduced by Lord Wynne-Jones in 1973. It passed its second reading though opposed by the Government and failed to reach the Statute Books because the Government fell. After three further attempts it was finally taken by the Government who brought in their own legislation in the House of Lords. Conservation Groups (World Wildlife Fund, Fauna Preservation Society, Friends of the Earth, International Council for Bird Preservation, Royal Society for the Protection of Birds, The Royal Society for the Prevention of Cruelty to Animals) worked together and met the Department of the Environment to agree on a form the Bill should take, because they were most unhappy with some aspects—in particular, the control of animal products. They felt that all products of rare and vulnerable species should be controlled. Another problem is the way the controls are to be brought in, for it gives traders a warning period in which they can stock-pile supplies. For example, from January–September 1975, 27,000 ocelot skins were imported, but in the three months October–December it rose to 76,000 skins because the traders knew (as was affirmed by their journal) that controls were being brought in on 1st January 1976.

The Conservation Bodies disagreed with the Bill in that they felt licences should not be issued for rare animals for the purpose of scientific research unless no other species can be used and the rare species could not be bred in captivity.

As it now stands the Endangered Species (Import, Export) Act 1976 which received Royal Assent in February 1977 prohibits the import or export of all animal products and plants not in its schedule without a licence which is issued by the Department of the Environment, on advice from the appropriate scientific authority—the schedules having been reversed so that only small numbers of animals listed on the schedule can be imported freely. Pressure from conservationists has obliged the authorities to permit private individuals to prosecute anyone exhibiting or offering for sale live animals or products therefrom that have been illegally imported.

Friends of the Earth propose to publish a manual on the status of mammals and reptiles most threatened by trade so as to reduce this pressure. It will also emphasize the problems and irregularities. Since June 1976 in spite of all the discussions, only skins of otters, colobus monkeys and turtles and a few made-up goods have been added to the list.

It is early days yet but there appears to be goodwill in many quarters for protecting Endangered Animals and Plants yet still disagreement as to how this should be done.

1976 also saw the Dangerous Wild Animals Bill become law. This Act is simply a control on the purchase and ownership of animals that might prove dangerous to man if they escaped, but it does serve a dual purpose. Before a licence is issued to prospective owners they have to prove that the accommodation provided is not only secure, but appropriate to the species concerned. They must also show that they have a basic knowledge of the species they desire to own and adequate funds to ensure its well-being whilst in their care. To many this might seem a small step in controlling the abuse of wild animals, yet prior to this legislation the private ownership of exotic beasts was quite incredible. We are personally acquainted with so many cases that it would be absolutely impossible to detail more than a few: a lion kept in a van in Woking; a Patas monkey that was shut in a kitchen pantry for 17 years; a puma that lived in a suburban garage in northern England; a collection of large cats and monkeys that almost constituted a menagerie, permanently housed in an old barn. Under earlier legislation it would have been almost impossible to take any effective action against these cases (it is virtually impossible to define cruelty in a British Court of Law), but now it is only necessary to provide evidence that the Act is contravened to produce results. Welcome as these measures are, they are long overdue, and our own personal enthusiasm is tempered by Dickens' line, 'The law is a ass—a idiot', for to date it has proved itself so with regard to wildlife protection.

Horrifying as the recent importation figures are, it is abundantly clear that the zoo trade's involvement is infinitesimal compared with the ravages of the pet, fur and raw-material importers of wildlife. But because of the zoological organizations' avowed obligation to wild animals, this alone is excessive and unacceptable. The most positive step that could be taken in this field would be to clamp down hard on the animal-dealer, or even eradicate him altogether.

It is difficult to ascertain the number of dealers in Britain, but it must run into hundreds. Just a brief glance at the classified advertisements in magazines like *Fur and Feather* or *Cage and Aviary Birds* reveals the volume and extent of the trade. Most of these are small-time operators for whom animal-dealing is only a side-line. They are a varied and colourful bunch nonetheless. One fellow we encountered made his living from selling used cars, but also ran a profitable line in exotic beasts. We visited his premises—a large, decaying farmhouse surrounded by rusty motor cars and lorries—not so very long ago, and were led into a ramshackle out-house adjoining the main building. Inside we were shown the merchandise for sale, which included a puma, two green monkeys, a Patas monkey, two macaws, several boa constrictors, pythons, rare Siamese crocodiles, a Nile crocodile and an anaconda. He refused to divulge the source of these animals, but to our certain knowledge the monkeys and puma had been sold to him by zoos; presumably the reptiles had been imported direct. Under the restrictions of the Dangerous Wild Animals Bill his activities should be largely brought to a halt, but at that time there was little doubt that he would have sold to anyone who wanted to buy. Outlets like this allowed animals like monkeys and big cats to become owned by totally inexperienced people. At the time of our visit to him it was strongly rumoured that certain safari parks were selling off excess stocks of lion cubs to private owners. When we tackled him with this he just said, 'If you want lions I can get them, but don't ask questions about their source'.

Bird-dealers are to be found in the greatest profusion, and so

*A selection of animals seen on dealers' lists: June 1977*

| | | | | |
|---|---|---|---|---|
| 1. | 1–2 Parma wallaby | (*Macropus parma*) | born 1976 | £520 trio |
| 2. | 1–0 Giant bushbaby | (*Galago crassicaudatus*) | born 1976 | £80 |
| 3. | 3–3 Dourocouli | (*Aotus trivirgatus*) | young adult | £300 pair |
| 4. | 3–0 White-throated capuchin | (*Cebus capucinus*) | born 1976 | £120 each |
| 5. | 3–2 Brown capuchin | (*Cebus apella*) | young adult | £210 pair |
| 6. | 6–6 Spider monkey | (*Ateles geoffroyi*) | young adults | £325 pair |
| 7. | 3–3 Black-headed squirrel monkey | (*Saimiri sciureus*) | born 1976 | £190 pair |
| 8. | 2–2 Silvery marmoset | (*Callithrix argentata*) | young adult | £330 pair |
| 9. | 1–1 White-lipped tamarin | (*Saguinus labiatus thomasi*) | young adult | £600 pair |
| 10. | 1–0 Gorilla | (*Gorilla gorilla*) | been in British zoo for past seven years, aged 9 years | details on request |
| 11. | 1–1 Chimpanzee | (*Pan troglodytes*) | approx. 1 year old | £1,600 pair |
| 12. | 1–2 Mayotte lemur | (*Lemur mayotensus*) | continental zoo born '73/'76 | £1,600 trio |
| 13. | 1–2 Giraffe, Masai, reticulated | (*Giraffa camelopardalis*) | British zoo born, 1972, 1975 & 1976 | £2,750 each |
| 14. | 1–0 Giraffe, Masai, reticulated | (*Giraffa camelopardalis*) | 9 months | £2,500 |
| 15. | 2–2 Gemsbok | (*Oryx gazella*) | continental zoo born '75 & '76 | £3,300 pair |
| 16. | 1–2 Jaguar | (*Panthera onca*) | born spring 1976 | £1,350 trio |
| 17. | 1–2 Siberian tiger | (*Panthera t. altaica*) | born 1st wk. Dec. | £700 each |
| 18. | 1–3 Indian tiger | (*Panthera tigris*) | born begin. Feb. | £650 each |
| 19. | 1–1 Indian tiger | (*Panthera tigris*) | 18 months old | £1,300 pair |
| 20. | 0–1 Sumatran tiger | (*Panthera t. sumatrae*) | born 1975 | £775 each |
| 21. | 1–1 Serval | (*Felis serval*) | born Dec. 1976 | £425 pair |
| 22. | 2–2 Black-back jackal | (*Canis mesomelas*) | born 1976 | £480 pair |
| 23. | 1–1 Bat-ear fox | (*Otocyon megalotic*) | young adult | £620 pair |
| 24. | 2–2 European otter | (*Lutra lutra*) | born 1975 | £920 pair |
| 25. | 1–1 Small-claw otter | (*Amblonyx cinerea*) | young adult | £690 pair |
| 26. | 1–1 African wildcat | (*Felis libyca*) | born 1976 | £520 pair |
| 27. | 2–4 Cape hyrax | (*Procavia capensis*) | young adult | £240 pair |
| 28. | 2 Emu | (*Dromiceus novaehollandiae*) | born 1976 | £800 pair |
| 29. | 1–1 Black-neck stork | (*Xenorhynchus asiaticus*) | young adult | £580 pair |
| 30. | 1–1 Wattled crane | (*Bugeranus carunculatus*) | young adult | £2,400 pair |
| 31. | 2–2 Andean cock-of-the-rock | (*Rupicola peruviana*) | young adult | £800 pair |
| 32. | 1–1 Cuban crocodile | (*Crocodylus rhombifer*) | 6 feet | £800 pair |
| 33. | 1–1 Reticulated python | (*Python reticulatus*) | 8/10 feet | £200 pair |
| 34. | 1–1 Yellow anaconda | (*Eunectes notaeus*) | 4/5 feet | £120 pair |
| 35. | 7 Malayan monitor | (*Varanus salvator*) | 6 feet 30 kilo | £250 each |

The figure before the dash represents male specimens, the figure following, female.

Numbers 1–8, 14, 21, 22, 23, 25, 26—all in quarantine.

Note the crossbred Masai/reticulated giraffes, numbers 13 & 14.

Several animals are clearly related by birth, for example the trio of jaguars born in 1976 are doubtless a male and its two sisters. Inter-breeding is quite commonplace. It is also disturbing to see rare species like the gorilla and Sumatran tiger on open sale: this means that any zoo, irrespective of its capabilities, can buy them. The only safeguard against this is the dealer's own integrity, and we have little faith in that!

are importers of reptiles and amphibians. The reasons for this are depressingly obvious. Import restrictions are weak. As we have seen, various Colombian reptiles were imported in 1976 despite the fact that they are totally protected in their country of origin. As for birds, these usually only have to serve a 30-day quarantine, after which they can be freely sold. Such trade is not confined to zoos; many of these animals end up in pet shops. Birds, reptiles and amphibians are the most susceptible groups when it comes to capture, transportation and the rigours of an alien climate. Their mortality rate is far above average for wild creatures. The trade is knowingly based on these factors. If 90 per cent of the tortoises imported into the United Kingdom did not die every year, the incentive for importing further shipments would cease, simply because the demand from pet purchasers would be reduced.

The same maxim applies to creatures like finches. Every year they are imported in their hundreds of thousands; every winter they die in equal numbers. Several of the less desirable zoos and wildlife parks are in the habit of buying 'job lots' of small birds each spring to give a splash of colour to the bird section. As the cages are only suitable for exhibition, the majority of birds die off in the winter, so a fresh batch is purchased for the following season. Much of the dealer's business is founded on this basis of supply and consumption.

One typical, medium-sized dealer we visited specialized in stock like reptiles and birds. His establishment appeared well-run and clean, and the majority of animals kept there seemed in good condition. A long, heated building had been utilized for a reptile/amphibian house and contained rows and rows of tanks filled with snakes and lizards. At the end of the building several heated pools held a variety of terrapins, crocodiles, alligators and caimans, and large sand-filled pits housed a wide selection of tortoises. Two adjoining buildings (also heated) were lined with stainless steel cages that were overflowing with thousands of brightly coloured tropical birds: glossy starlings, barbets, toucans, parakeets, manakins, waxbills, flycatchers—boxed together by the hundred.

Nearly all of these creatures would be destined for an early grave at the hands of private owners who lack the experience to care for them, or in sub-standard zoos or wildlife parks where amateurism is equally commonplace. Viewed in this realistic light, the thousands of creatures in the rows of tanks and cages gave a very unpleasant indication of the scale of wildlife consumption in Britain.

Major animal-dealers in Britain are few, there are less than half a dozen in permanent operation. This is understandable, as large-scale animal-dealing requires a considerable amount of capital to start with, and nerves of steel thereafter, for even a moderately sized import of large mammals can involve a substantial investment. The problem with these enterprises is basically the method of operation: animals must first be brought into Britain and then a purchaser must be found. This means that far too many specimens remain at the dealers waiting to be bought, when they should be settling in their permanent captive environment.

Earlier we mentioned Gerald Durrell's expedition to collect Colobus monkeys for the Jersey Wildlife Preservation Trust. In that instance the animals were captured and transported by one group of people and on arrival at the zoo were settled into cages that were to remain their permanent home; since then a healthy, thriving breeding group has been established. By comparison, a British dealer also imported some Colobus in the early 1970s, with catastrophic results. One of the staff who drove to the airport to collect them said it was, apparently, a very cold night and the heater in the van was not working properly, so by the time the monkeys arrived at the dealers several were suffering from exposure. This was a bad start to settling in one of the more delicate species of primate, and it seems that they never really recovered from the experience, for very shortly they all died. Various medical reasons have been given to account for the deaths, from severe enteritis to hepatitis. But the fact remains that they all died, whereas none of the Jersey import did. Considering the past record of both importers, the reasons seem self-evident. In the case of the Jersey import, the animals were moved into specially designed cages on arrival in Europe and the settling process started immediately; in the latter instance the Colobus were purchased from some obscure dealer in West Africa, flown into Britain unattended and taken to an establishment where the accommodation was inferior.

Controversial as these criticisms are, they seem to be applicable in the general sense to nearly all operations carried out solely by a dealer, and it is our contention that this is because the wild specimen is viewed only as an object for retail, instead of a creature composed of flesh and blood. Naturally dealers do their best to see that the animals survive, but by cutting corners and making economies in order to keep the operation financially viable, much is left to chance. We have already mentioned inferior crates—a typical dealer's economy. There is also the practice of shipping animals unattended, which makes an obvious saving on fares and hotel bills, but it is often a reckless course of action. However, the greatest savings of all are made in the dealer's yard.

A large animal-dealer needs an incredible variety of cages and enclosures to accommodate the thousands of scaled, furred and feathered creatures which are destined to pass through his premises each year. Heated rooms for reptiles, small birds and delicate mammals; aviaries and pools for birds; strong cages for the larger carnivores; and, equally, strong pens and shelters for big animals like elephant or bison. Creatures like sea-lions need water, monkeys, well-branched cages and heated sleeping rooms. Yet however well planned the complex is, this transit accommodation is never wholly satisfactory as it cannot be anything more than a basic module designed to house a type of animal, rather than a specific one. Cages always tend to be small and unimaginative, and, in addition, because of the dealers' economies, they are frequently badly overcrowded.

Visiting dealers has always been an undertaking that we approach with awful feelings of gloom and foreboding, and regrettably this initial apprehension generally turns out to be more than justified. A visit made to a British dealer in the early

1970s provides a typical picture of the conditions attendant to this trade. On the day of our particular visit a young leopard (wearing a dog collar) was pacing restlessly up and down in the window, much to the delight of some young children standing in the street; but the animal was clearly nervous at being in so exposed an area and the sight did not give us much pleasure.

Inside, a confused spectacle greeted us. On one side was the usual array of pet foods, rubber bones, bird seed and doggie chocolates, on the other a variety of parrots and macaws in small cages fixed on to the walls. Beneath the birds were two wire cages, possibly 6ft by 4ft, within which stood two mournful-looking young Malayan tapirs. The male (they were a true pair) was covered in sores or abscesses and the hair on its body was missing in several areas; it looked very sick indeed. We were assured by the owner that the tapir was receiving veterinary treatment and he explained that he was very upset by its condition. He had purchased it in good faith and was horrified to find such a sickly animal when it was uncrated. We did not manage to extract the source of these two animals. Both were young and of almost the same age, but it is not conceivable that they were zoo-bred as there are only about 150 Malayan tapirs in the zoos of the world and only 8 births were registered during that year. Therefore these animals were caught and exported from the wild, where the species is already in severe decline because of the increased clearing of its forest habitat, and shipped off—not to a zoological organization that might endeavour to breed them, but to a middleman whose sole purpose was to make a profit. And for this reason alone they were condemned to suffer the restrictions of their wire cages until a suitable purchaser—one not put off by their poor condition—could be found.

In sharp contrast to these rarities, we found in the outside yard several species so common to the zoos and wildlife parks of the United Kingdom that it was difficult to foresee how the dealer would ever sell them. Pumas breed readily in captivity and, due entirely to the short-sightedness of zoo management in the zoo boom of the 'sixties when new zoo-owners sought animals that were easily obtained, cheap, yet still exotic enough to be considered 'crowd pullers', the British market is now flooded with unwanted cubs.

The same has happened with lions. Surplus stocks meant that they were suddenly available on the open market at a price within reach of the average citizen, so a new craze developed, and would probably have continued had not the Dangerous Wild Animals Bill been enforced. This dealer had unwisely purchased over a dozen zoo-bred pumas, which he had placed in cages in his back yard, and five lions which were housed in the only accommodation left: an unused room over the shop. Other common species that would prove difficult to sell included several Rhesus monkeys, some bears and baboons. Doubtless all these were destined for a long period of internment in their temporary cages.

Many Chilean flamingoes were also resident in the outside yard, although we were told that they would not be there for long as most of them had been sold and would be delivered to their new owners within a few days. The trade in flamingoes is a source of constant aggravation to us, mainly because few people seem concerned at the number imported each season, which is a clear indicator of the annual mortality rate and the number which die in transit or dealers' yards. They are a typical example of the practice of buying cheaply at source and revaluing to cover (or more than just cover) mortalities prior to delivery. On a good shipment, the dealer might expect to obtain Chileans at about £15 a bird, sometimes less. To this he must add the cost of crating, air-freight and feeding when they arrive in Britain—and deaths. In 1976 these birds were retailing at prices between £85 and £100 each in this country. It is difficult to determine exactly how much of this is actual profit, for much depends on how many birds there were in the original consignment (air-freight per head is obviously cheaper for a shipment of 500 birds than it would be for 50), and how many of the consignment die before reaching the final destination. Deaths are certainly accelerated by the practice of wrapping long-legged species like flamingoes in ladies' nylon stockings then placing these living bundles in long cardboard boxes similar to those used for shipping cut flowers. The stocking method is quite advantageous when used over short periods as the bird is restrained and protected at the same moment; it cannot fall over, kick or injure itself or its travelling companions, and will usually settle quietly within a few minutes of being secured. Over long periods however, this method can jeopardize the bird's life. The veins and arteries in its pencil-thin legs are understandably very small, so small that it takes only minor interference with the blood flow (as happens when they are wrapped in stockings) to damage the legs permanently, and a crippled flamingo is as good as dead. Properly constructed crates should be used. Indeed an excellent blueprint of a container for these birds can be found in the IATA handbook; but the stocking and cardboard box system is preferred by those who can get away with it because it is so cheap: the cost is negligible compared to a crate.

Of the flamingoes we saw in the yard, several were in a poor condition. Some hobbled on one foot, scarcely able to place the other on the ground, and three were lying down on wet concrete, looking very bedraggled. Later one of the zoo purchasers informed us that shortly after the arrival of his birds, one died. A post-mortem was hardly required as a cursory examination revealed a raw, weeping area on the breast which was soon confirmed as severe necrosis of the breast-bone (in other words, gangrene), most likely caused by lying on wet, cold concrete. Rather significantly as we left the yard we saw a pile of discarded nylons and long cardboard boxes that had been thrown into a corner!

If this dealer seems revolting, we have seen others as bad, and have received reports that hardly seem credible, although we know them to be true. An experienced zoo-keeper of our acquaintance strayed from the zoo world and spent several months working for a dealer. He maintained that never had he seen so much death in so short a space of time. He gave vivid descriptions of removing the dead crocodiles and caiman from the reptile pens, of dumping monkey carcases in the incinerator, of watching animals die from sheer neglect. Anxious to remain employed once he had joined another zoo, he refused to give evidence to an animal welfare organization, so the situation remained unresolved.

Perhaps, from the brief examples given, the reader can understand our anger and frustration. We have laws to protect Britain from infection and disease: some adequate, some not so good. We have laws intended to safeguard the future of wildlife, by restricting the trade in endangered species, and most of them are farcical. And we have the animal-dealer, who, despite the regulations that do exist, still manages to get hold of most of the wildlife entering the country and a fair amount of that which is already settled here. It is sad to know that even some of the best collections are prepared to sell surplus stock to dealers and are content to allow their animals to go to temporary accommodation to await a purchaser, all of which seems to imply that once an animal is sold its fate is of no further concern to them.

Contrary to the firm conviction of most zoos, dealers need not be the sole source of animal stocks. In fact dealers need not exist at all were it not for the blindness and stupidity of the zoos themselves, for the solution lies in their hands and has done so for years. If zoos worked together harmoniously, surplus stocks could easily be exchanged within the zoo network without the intervention of dealers, and on the occasions when the capture of wild animals was justified, it could be undertaken by the zoo authorities in close co-operation with the conservation bodies of the country of origin. Stated in this way the answer seems so absurdly straightforward it must be incomprehensible to most people that a central zoo agency does not exist. It does not because there would be little profit in it and its existence would bring to a halt the activities of a number of business enterprises. Attempts have been made to establish something on these lines. To date all have failed. The ex-Secretary/Director of Glasgow Zoo told us of his efforts to set up a central office of information on zoo stocks. The idea was that each zoo in the country would send up-to-date stock-lists to him and from these it would be possible to pair up (on paper) solitary specimens or surplus stocks and negotiate an arrangement between the owners. The response was so poor that he abandoned his efforts within only a few months.

'Surplus to requirements' and 'Animals wanted' lists are produced by organizations like the National Zoological Association and the Federation, but only include animals belonging to their members. Also the excellent magazine, *The International Zoo News*, regularly prints similar lists, yet these attempts to communicate are spasmodic and undisciplined. Disciplined organization is decidedly lacking in British zoos, and some show extreme reluctance at revealing births, deaths or even sexes of animals owned, which was proved by their refusal to co-operate with the Zoological Director of Jersey when he attempted a census of the great apes, or the failure of the proposed central agency for animal stocks. Many even decline to make annual returns to the *International Zoo Yearbook*.

Co-operation could achieve so much, resistance to it seems foolhardy. The only animal-dealer in Britain for whom we have regard actually works along the lines we are proposing. No dealer's yard exists; information on wild animals available for purchase and zoos' requirements is correlated in an office. Once a transaction has been agreed, the animal is crated and transported directly to the purchaser. A more accurate term for the dealer in this case might be broker, as effectively business is transacted on behalf of others without the need to resort to dealers' yards and temporary caging. If British zoos worked together and in the full interest of the animals upon which they depend so entirely for their existence, this method of trading could have become standard years ago. They might also have reached the stage of having a central zoo authority which could run a specially designed quarantine reserved exclusively for zoo specimens. With such an authority they could have become self-policing, ensuring that all animals went to places capable of caring for them, rather than to the highest bidder. But they have not even approached this stage yet. So whenever a flamingo chick emerges from its shell in a captive environment, the proud owner should in all moral conscience temper his delight with the thought of the countless birds that died to make the small miracle possible.

Chapter 5

# A living plan

Whether a collection of wild animals is designated a zoo, wildlife park or safari park, the methods of husbandry differ little and the secret of success in each case rests entirely on four factors: adequate finance, sound management, intelligent design and reliable and dedicated staff.

Finance is of primary importance and, in this area, British zoos are unique in that most are commercially owned. There are only a few municipal zoos, which are nearly all small, and a few Zoological Society zoos, which include the oldest in the country. This is contrary to the situation in the rest of the world, where most zoos are owned by municipal or federal authorities, with a much smaller number of commercial organizations in operation.

Most of the collections that have opened in Britain during the last 15 years have been under-capitalized from the onset, primarily because the financial backers or owners have only the vaguest idea of the cost involved and have assumed that the bulk of their money would be needed for the purchase of land and animals. In reality, these account for only a small proportion of the total investment required. Most of the initial capital expenditure must be spent on facilities that do little to enhance the appearance of the collection, but are essential if it is to operate smoothly. Drainage and power lines must be put in, perhaps woodland has to be cleared, paths or roadways laid, and a perimeter fence should be erected to surround the whole property. We say 'should', for many wildlife collections currently in operation do not have perimeter fences because they are deemed too costly to erect and not vital. This should not be tolerated. Every collection should be surrounded by a security fence for two reasons. It keeps out intruders and vandals, and it retains any animal that has chanced to escape from its cage or enclosure. In the event of an escape it is obviously preferable to know that the animal is still within the zoo grounds rather than roaming the open countryside. If the latter is the case, it often means that the escapee is shot to safeguard the public, or—if not found—it is almost certain to die of cold or starvation.

It is impossible to give a precise figure for the minimum investment needed to develop a collection, because of the rapid rate of inflation over the past few years, but the following figures calculated at the end of 1977 might be helpful as an approximate guide. A small specialist collection would need not less than £100,000 to £150,000; a medium-sized one something between £300,000 to £500,000 initial capital, and a major mainstream zoological centre could not be considered unless at least £2 to £3 million were available for development. To explain what happens when a collection is under-financed we will examine some cases of small zoos we visited in the early seventies.

One collection had been set up five years before by a wealthy landowner who had decided to turn the most attractive part of his estates into a recreation area, and he had certainly chosen a beautiful site. The main section of woodland and grass sloped gently down to a lake containing waterfowl. As a setting for a wildlife collection it was idyllic.

When it first opened it was essentially a bird collection. The lake was stocked with a variety of waterfowl and a flock of free-flying Canada geese soon settled there amongst the other birds. Aviaries were erected in the main area of the park and filled with colourful foreign species. A number of birds of prey were purchased and some paddocks were built to contain a few deer. It was a modest little venture that was pleasing to the eye without causing concern to those involved in animal welfare, or at least this was so in the very beginning.

By the time of our visits to the park it had undergone a dramatic change. The original construction work had been executed as cheaply as possible (always a false economy), and many of the cages and enclosures were already dilapidated, although only five years old.

A large, and rather attractive, walk-through aviary built entirely of timber and thin wire had started to rot. The main supports were giving way and the wire top had constantly to be repaired to prevent the smaller birds escaping. Even worse, the builders had neglected to lay any footings, or to underseal the aviary with wire to prevent rodents getting in, so the place was overrun with rats in search of an easy meal. Old seed and fruit were their main interests but they were equally partial to any birds unwary enough to land within their reach, and several specimens met their untimely death in this manner.

The deer paddocks had been made from an assortment of material. Rough hewn poles were driven into the earth to provide the supports for chain-link fencing on one side and on the other some rather old, misshapen sheep-netting that had been found elsewhere on the estate. Interior furnishings included a lavatory cistern which served as the water trough, while the deer-houses were nothing more than curved corrugated metal roofs (like those often seen on the huts of construction gangs) placed on the ground and filled with bedding.

A llama, in a similarly constructed paddock nearby, had a shelter made from a section of lapboard fencing and several ancient doors nailed together to form three walls and a rather leaky roof.

As for those invisible factors that account for so much of the early capital investment, they did not exist. Two taps, both near the main yard, were the only source of water in the whole 30-acre park, so all water had to be taken to the animals in four-gallon tanks balanced on a wheelbarrow—a very slow, laborious task. Power lines ran to an existing building that constituted the food preparation and store rooms, also to two wooden buildings alongside and the cafeteria and gift-shop, but elsewhere there was no source of energy available. This meant that in an emergency, if an animal was sick or injured for instance, it was impossible to get light and heating to it, and as the park had no sick-bay it was equally impossible to move it to a place of warmth where it might be treated. This leads us to another facility the park did not have—a place for veterinary examinations.

Another common fault found at another collection involved the food store and food preparation rooms. It is an accepted rule that these places must be made rodent-proof and that all working surfaces and containers should be of materials that are easily cleaned, but here a rickety old kitchen table sufficed for a working surface and the building was full of rats. The problem of rodents is a headache for every zoo director. It is almost inevitable that they will invade an animal collection and one has to take every precaution possible to stop them. Rats and mice will spread disease, contaminate foodstuffs with faeces and urine and, as was the case in the walk-through aviary, even maim or kill stock. In this collection no precautions had been taken, save for the monthly visit of a rodent control officer, who gingerly spread his poisons in places where it was impossible to injure the park's inhabitants. Understandably, the war against the rats was more symbolic than effective.

Here, the one major investment made when the park was planned was a tarmac road. But other essentials like drainage were omitted completely, whilst the perimeter fence surrounded just one part of the establishment, and as it was only four feet high with a single strand of barbed wire running along the top it can hardly be considered a security measure.

All this sounds depressing enough but often the situation is further aggravated by an owner's development programme, for as time passes he starts to veer from the earlier concept of a simple collection consisting of birds, a few deer and a pets' corner, to purchasing more demanding species. In this expansion of the park's style of exhibition, they can be severely hampered by their total ignorance of wild animals, and indeed a tragically funny story exists regarding such an incident.

It seems that while wandering around a dealer's yard a zoo owner saw an adult bull wildebeeste which was being offered at a ridiculously low price. Always a man to seize a bargain, and possibly a little inspired by a recent visit to a safari park where these animals present a fine spectacle grazing on the open park-land, he bought it. Whether he intended to obtain some females for it later or was content to settle for a solitary bull we shall never know, for he had purchased the animal without making provision for housing it. With a bravado only amateurs dare display, he announced the coming of the beast to his staff a few days prior to its arrival, and told them that it would be temporarily housed in an unused bull-pen until a paddock could be erected. Apparently he also added that he could not foresee any problems as the animal was half the size of his cattle, and in his time he had handled bulls three times as large. But wildebeestes are not cattle, they are antelopes, and fairly aggressive antelopes too if the occasion merits it; and this bull obviously thought his situation more than justified a bit of hostility. Once they got him into the bull-pen and closed the heavy doors, there was no way to get back in to feed and water. Each time they tried to open the door the bull charged it, threatening to injure staff and escape. Cleaning out or replacing bedding was quite out of the question and to make matters worse there was no way to get him out again, at least not with comparative safety. This incredible farce lasted some days. The keepers devised a way of getting food and water in for the animal (by spilling it in panic mostly), the owner lost his nerve, abandoned paddock-building and asked the dealer to take it back. In the end this was agreed upon and the poor creature was drugged by a local vet, crated and returned from whence it came.

At another collection, the monkey range was made from second-hand timber huts, purchased at an auction, blocked up against what had formerly been aviaries, two of which had been strengthened with strong wire but the rest remained as they were when the occupants were birds. Naturally there were escapes, for each cage was a patchwork of repairs. Three adult macaques escaped into the surrounding countryside, and as the park had neither an adequate security fence nor suitable equipment for capture, the staff were obliged to shoot the two largest specimens to safeguard the residents of the local village. It was a killing that should never have taken place. After this happened, the owner rebuilt the outside runs, although he would not permit the demolition of the wooden huts. The new cages are a great improvement on their forebears. They are almost 16 feet high where the old ones were 6, and they are constructed on a concrete base from strong timber and heavy-gauge wire-mesh, but the sleeping quarters are still those ugly, draughty old sheds, so the overall improvement is marginal.

We have said that adequate finance backed by sound management is essential if a zoo is to operate successfully, but this park lacked both and thus a chain-reaction was set up that destroyed any hope of worthwhile designing or the appointment of dedicated staff.

The problems of running a medium-sized park which lacks suitable back-up facilities such as water and power are daunting, although they are not insurmountable if there are enough staff and suitable funds to plan a programme of redevelopment. However, we heard of an owner who refused to employ more than three keepers who, despite the curator's additional assistance on routine feeding and cleaning duties, were hard pushed to get through their daily tasks, let alone the extra work he wished them to do. Beyond caring for the animals they were expected to undertake fence repairs, paint, mow the grass, collect litter, and do all the building work in the park. The sheer impossibility of accomplishing all this accounted for most of their lack of enthusiasm, their low wages for the remainder.

When it was agreed that a new exhibit should be developed, or an old one replaced, the owner rarely ordered the materials from a wholesaler; instead he embarked upon a grand tour of the local scrapyards. Periodically, vans would arrive laden with old timber, galvanized tanks, nails, screws and rolls of battered wire. From this assortment of old junk, the staff were expected to 'design' and erect a cage suitable for an animal to live in.

One of the largest purchases he made was of several thousand plywood boxes from a frozen food company. These were cut into separate panels and used for building work. Fifty per cent of the animals in the collection lived in accommodation made from these, and to convert a timber building into a modest reptile house, the main materials provided were several sheets of plate glass from some recently demolished shops, timber of various shapes and sizes from which nails and screws had to be removed prior to use, dozens of kerb-stones taken from a nearby lay-by, and, of course, the plywood boxes.

In view of the extremely limited standards within this type of park, certain species, like tiger, should never be considered for exhibition. For example, in one park, a pair of Bengal tiger cubs (almost certainly brother and sister) were obtained, and after a couple of years the male died. Solitary specimens always cause concern, and rare solitary specimens like tigers the more so. It was therefore suggested that it would be better for the park, and the tiger, if a new home with a suitable mate were found for the lone female. The sole reason for the owner's rejection of this suggestion was that the latest edition of the guide-book had just been printed with both photographs and text about the tiger! To leave a healthy young female on its own would have been scandalous and the only alternative was to obtain a mate of a suitable age and attempt to get them to breed.

The obstacles to this endeavour were simple: firstly, to find the right male and pay for it (tigers at this time cost £800–£1,000 each); secondly, to finance the adaptation of the existing accommodation so that it was both suitable and safe for the introduction of adult tigers.

The tiger enclosure was described in the guide-book as, 'designed to give as natural a setting as possible and to offer privacy to these animals so that it is hoped in time breeding will take place'. In fact the enclosure was built in a small copse without regard for the biological requirements of the species. It was made from telegraph poles and chain-link, with a large wooden shed containing two dens as a sleeping area. Under such conditions it would have been exceedingly dangerous to introduce a strange male to the resident female, for if they fought it would have been impossible to separate them. Ideally, the existing exhibit should have been enlarged by having a second, bigger enclosure erected alongside the first with several communicating tunnels (that could be shut off with slides) running between the two. This would make it possible to keep both tigers separated without depriving either of fresh air, sunlight or exercise, whilst allowing them to see each other and get acquainted. It would also give the staff time to analyse the reactions and gauge the possible outcome of an introduction.

It is often innocently imagined that breeding wild animals requires no more effort than putting together a male and female (some zoos do not even get this far). However, it generally takes a lot more careful consideration and planning. In the first place, the prospective bride and groom might loathe each other on sight and if this is so, no amount of human interference is going to change their minds, (the Jersey gorillas provide such a case). Secondly, one has to remember that if two members of the same species meet in the wild and fight, the loser is almost always given an opportunity to escape, but in captivity physical barriers block off any escape route. This can easily result in the vanquished animal being killed by the more powerful one. Such considerations must temper all attempts to introduce wild animals with a great degree of caution, with perfectionist killers like the tiger, the caution is stretched to its limits. Every precaution that can be taken, must be, and one must be reasonably certain that if serious fighting starts it will be possible to separate the antagonists.

The park owner, despite his reluctance to spend a sizeable sum on the operation, agreed that the second enclosure was necessary if a mate were to be obtained, and so the search for a male began. Within a very short time a suitable animal was found. This was an adult male which had successfully sired a number of cubs and was without a mate at a private zoo. The owner was prepared to place the tiger on permanent loan to a collection that had suitable accommodation and a mate for it. The only blemish on his character was that he had killed one of his females. This did not mean that he was an abnormally aggressive creature, but it did emphasize the need for care when introducing him to a new mate. If that was accomplished successfully it should only have been a matter of time before the pair reproduced.

Upon the tiger's owner generously agreeing to it being deposited with the park's female in the hope that they would breed, the curator went about the business of ordering building materials for the second enclosure, and organizing temporary facilities for the arrival of the male. For a strictly limited period the male would occupy the second den and use the outside enclosure on a rota system with the female, one outside in the mornings, the other in the afternoons. This was not ideal but it had the hidden advantage of allowing both specimens to become familiar with their partner's sounds and scents before actually facing each other through a wire barrier. The situation was not intended to last long anyway.

But when the materials for construction arrived, the owner said that he did not intend to employ workmen to erect the pen as it would be too expensive. He expected the staff to undertake the work in addition to their usual tasks, though none of the staff was a qualified builder or carpenter. Unfortunately, it took so long that the tiger was eventually returned to its original owner and the planned mating never took place.

Should the reader think that this collection is in any way exceptional, we hasten to point out that we know of many similar. Most are relatively new developments, ten years or less in operation, and owned by people with an appreciation of the wild animal's magnetic appeal to the general public, but with little or no knowledge of the requirements of their stock or the purposes of zoological exhibition. This blocks progress in zoos from design of the captive environment to standards of training for animal attendants. In fact hardly any zoo keepers in Britain receive any training at all.

Although few people have the opportunity to see behind the scenes when visiting zoos and therefore cannot know whether there is a veterinary complex or not, or whether the food store is clean or rat-infested, it is still possible to form an impression of the validity of the project by looking for certain tell-tale signs. The most obvious of these is the condition of the animals. Do they have sleek coats and bright eyes, or do they appear out of condition? Do they have access to clean water and dry, warm shelter? But these are only very basic points, they do not always tell the full story. Zoological exhibition is an art. It is the art of creating for an animal a functionally appropriate environment which is secure without being obtrusive. A good exhibit is one that is designed solely for the animal's benefit but still manages to pass on to the spectator a sense of the well-being of the inhabitants. Carl Hagenbeck realized this. But do many of our zoos today?

Run-of-the-mill zoos with limited ideas or objectives can be thoroughly monotonous places to visit, and, when faced with cage after identical cage, even the most devoted zoo-goer can become decidedly bored, irrespective of the wonders contained within them. Furthermore, as we stated earlier, one visit to an animal collection is usually sufficient to obviate the need to visit any others, for the simple reason that they are often so alike in content and style. The same species are shown in collection after collection, often in the same type of cages or enclosures. Even the guide-books all look alike.

A good collection needs, as well as adequate finance, management capable of designing, or of advising designers and architects on how best to build the artificial environments in which the animals have to live. The collection we have just mentioned suffered from acute financial starvation and therefore became a negative collection without merit or value but, had suitable money been made available, it would still have required a considerable amount of skill and experience to develop the collection wisely. It is experience and imagination that is totally lacking in most zoo development in Britain today. Many of the new directors do not plan their exhibits (primarily because they could not if they tried); they merely visit other collections and crib whatever ideas capture their imagination, and these invariably lose something in translation.

A highly satisfactory way of exhibiting otters, for example, is to develop a sizeable enclosure with a pool and sleeping dens. To provide added stimulus for both animals and public, construct a

9 Orang-utan with infant at Jersey Wildlife Preservation Trust

10 Examples of bad housing in zoos

a. This Mona monkey and white-collared Mangabey have no outside run

b. Hippopotamus in a cage, in a zoo in South Africa

c. This rare Parma wallaby is hardly protected in the snow

d. Is this the way to keep a very rare and expensive Bongo?

11 Burchells' zebra with wildebeeste at Gorongoza National Park, Mozambique

12a. Przewalski's horses. These rare animals have been successfully bred in ma collections, including Whipsnade, Marwell and Howletts

b. Part of the spectacular breeding group of the rare Scimitar-horned Oryx, Marwell Zoological Park

large glass tank on elevated ground and connect tank and pool by a waterfall. A simple pumping system is sufficient to recycle water from the pool to the tank, so the cost of the exhibit can be kept at a reasonable figure, whilst the end-product proves an admirable home for otters and provides a pleasing spectacle to the visitor. These charming creatures can be then seen swimming underwater, tumbling down the waterfall to dive into the pond, or romping across the grass. This idea has been adopted by several collections, although not always with much intelligence. Thus one finds smaller and smaller tanks, and pools and enclosures that reflect little of the sophistication of the original concept.

An architect cannot be expected to know all about wildlife and its behaviour; he relies entirely on the brief given to him by the zoo management. However, if this is comprehensive enough, a competent architect should be able to design accommodation suitable for virtually any species. In the past, few biological or ethological considerations would have been included in this briefing, mainly because few were known, and most buildings were erected more for their imposing design than their functional attributes. Examples of this kind include Dudley Zoo's polar bear exhibit: a construction entirely of reinforced concrete which represents the style of architecture so common between the wars—ultra modern, highly technical—yet to the observer this unrelieved expanse of concrete is monotonous and unpleasing. It is in reality merely a twentieth-century version of the Victorian bear-pit. It was designed by the Tecton group of architects, who were responsible for a number of zoo constructions during the 'thirties, including a horrible gorilla house at London Zoo, which is thankfully no longer in use, and the famous penguin pool at the same place. Although quite unsuited to the requirements of penguins, because of its unique design it has been placed under a preservation order, and so London is obliged to retain it. There are literally hundreds of architectural monstrosities spread through the world's zoos, places where the environmental features have been allowed to dominate the species exhibited, and new examples still seem to crop up each year.

The failure of much zoo architecture surely rests on the lack of flexibility involved. What use is a million-pound building which is technically difficult to operate and maintain, which falls short of the ideal for the species contained within it, and which is bound to be classed as old-fashioned within a couple of decades? Yet a number of the major collections have sunk considerable sums of money into vast buildings which, when completed, have the proportions of cathedrals and the interiors of public lavatories. Undeniably such criticisms rest almost entirely on the individual's reaction to a particular construction. For instance, of the more recent developments at London Zoo—Snowdon's Aviary, the elephant-house, the primate and carnivore ranges—only the latter strikes us as being in any way exceptional, but many zoologists and architects would strongly disagree with this. Our basic criticisms rest on the idea that better (from the animals' point of view) could have been achieved at far less cost.

Let us take London's monolithic elephant pavilion. A leading authority on zoo architecture has the following to say about it:

> While this general plan springs from functional requirements, and is related to a circular food store and preparation area in the basement, two visual analogies have been noted and emphasized. Internally the low and dark central area, with its radial arrangements of timber joists, gives a forest effect, and this is enhanced by the animals being placed in bright, top lit clearings. The light to the animal areas trickles down from tall lantern lights dominating the exterior of the building, and these suggest a close visual parallel to a group of elephants around a watering place. Externally, too, the scale is elephantine and the ribbed concrete walls match the rough cast skin of elephants and rhinos.

The scale of the building is certainly elephantine, for there is as much of it going in the upward direction as there is at ground level; were elephants and rhinos arboreal, its proportions would be ideal. As they do not climb, however, the immense height of the building simply evidences a colossal waste of highly expensive building materials and as London hardly manages to make ends meet financially, this must be considered reprehensible. Like so much of zoo architecture the building seems to reflect the almost universal self-consciousness of the architect, an earnest belief that the quality of man's handiwork can surpass that of nature. References to radial timber joists giving a forest effect are ambiguous to say the least; they are also typical of the conceit of many designers.

One Society zoo, Chester, was at one time much criticized for some of its buildings which were considered 'un-architectural' and factory-like, but Chester is the one major zoo in Britain that has retained a greater degree of flexibility than any other. As knowledge of a species grew, so its accommodation was altered to suit the newly-found requirements. (Chester's letterhead actually displays a beaver beneath which are the words 'Always Building'.) Thus, today one finds that the second largest collection in Britain is also one of the most technically up-to-date. Although we do not personally like the large mainstream collections, because we consider that the concept is so totally out-of-date, we have to admit that the feeling of spaciousness and the delightful setting at Chester do admirable justice to what is a most important collection. It is a beautiful zoo by any standards and well worth a visit.

The architect or designer has three basic points to consider before embarking on a zoological development: the requirements of the species to be catered for, the security of the finished product and the ultimate presentation to the public; and the points should be considered in that order, Above all else, what is best for the animal must be the main thought behind the design, and this is followed closely by the security aspect of the cage or enclosure, for if an animal escapes it is bad for all concerned. Secure accommodation depends mainly on the use of sound materials, but equally these materials have to be used in a sensitive and intelligent way.

By way of illustration let us use the previously mentioned

escape of macaques at the down-at-heel zoo. In this case it is not just a question of how they got out of their cage, but why. We know that they escaped because the wire was too weak for primates of this size and strength, but it is significant that they devoted so much time and energy on tearing at the wire in the first place. The reason was straightforward enough. Inside the cage there were no branches for climbing upon, no rocks, in fact nothing in the way of occupational diversion for the animals. It is therefore hardly surprising that these intelligent mammals turned their attention to the one and only thing left to play with—the wire. Had their living conditions been of good quality, it is quite likely, despite the insecurity of the cage, that they would never have bothered with the wire at all. Security is thus a combination of environmental quality and sound structure. Macaques do not require cages of concrete and iron bars, but it is prudent to use heavy-gauge wire and humane to provide plenty of interest within the cage to excite their day-to-day existence; and thus escape becomes unlikely and unnecessary.

It is possible to emphasize this point even further with two distinctly different reactions to escape. Some years ago one of us worked in a zoo in southern England that had a pair of chimpanzees who were inveterate escape artists and, as they were both adult specimens, these occasional excursions into the zoo grounds were decidedly unamusing. Everyone, even the most experienced members of staff, felt extremely nervous when faced with the formidable task of apprehending these two potentially dangerous apes.

Their cage could not be described as bad, just unsuitable. It had been built when they were youngsters, specifically for small chimpanzees, and in time they just outgrew it. A new chimpanzee house was planned and, just as the building contractors began laying the foundations, the male chimp discovered a way of dismantling his cage and from then on all the staff had to be on the alert for his break-out attempts. The method of escape was delightfully simple, the cage was made from chain-link and the chimp had found that if he could locate a weak point in the wire, and break it, it was then easy to unthread the chain-link until a sizeable hole appeared. It cannot be denied that the world outside abounded with more objects of fun than did his cage, therefore it was understandable that he occasionally chose to go for a walk. But when provided with the attention of his keepers and things to play with, it never occurred to him to start demolishing the wire. Only when left alone and unoccupied did the chimps escape. The worst occasion was one Christmas Day when the staff were sitting down to dinner and the zoo was unusually quiet and empty. The festive meal was suddenly interrupted by the forbidding cry, 'The chimps are out!' When moved into their new house the question of escape no longer arose, for the accommodation was not only far more secure, but environmentally more satisfying.

The second example concerns a tiger, and the incident provides a strong case for the significance of environmental conditions. The animal was kept in a large cage that was furnished with a pool for bathing in, a stout platform, a den and other facilities that are known to be popular with tigers. As far as could be ascertained the tiger seemed more than content with his lot, but no one could have foreseen how dramatically this apparent contentment would be tested. On the day in question the writer was in a great hurry to complete some routine duties which included cleaning and refilling the tiger's pool. This he did and then opened the slide to allow the animal access to its cage, but in the rush he had failed to close the doors to the outside enclosure. It was an inexcusable mistake, for it is an unwritten law known to all who work with animals that one never rushes about, also that locks are always double-checked. However, the error had been made—a few feet in front of him and, on the wrong side of the wire, stood a large male tiger. Thankfully the animal was known to be friendly, in fact he was as gentle as a lamb; but the problem was not one of his tractability, it was how to get him back into his cage. Incredible as it may seem, the problem was resolved by the tiger itself. He wandered about smelling the interesting objects lying around: wheelbarrows, sacks, brooms. He even went over to a cage containing a pair of leopards but clearly disliked their spitting and snarling, and then, as casually as he had walked out, the animal re-entered his cage and sank blissfully into the crystal clear water of the freshly cleaned pool. The doors were quickly secured and the writer retired, somewhat ruffled, for a strong drink.

Some obvious conclusions can be drawn. First, cages or enclosures must be constructed from materials which are strong enough to contain the animals even under extreme stress. Secondly, but even more important, the environment should be satisfying enough to make escape irrelevant from the animal's point of view. It is the distinction between the psychology of restraint and the practice of incarceration. Bearing these points in mind, and also our rapidly increasing knowledge about keeping wild animals in captivity, the zoo architect should proceed cautiously, considering the variable reactions found between species, even within a species, and remembering that often designs are outdated while still on the drawing-board because of new discoveries made in wildlife management. Trees, grass, earth and water must be used as skilfully as concrete, bricks and glass. The end-product is supposed to be sympathetic to both the physiological and psychological needs of the animal, while allowing the observer, *Homo sapiens*, a feeling of being an integral part of the whole.

The monolithic is not necessarily the best. We can cite the examples of the primate cages at Twycross Zoo in Warwickshire and the ungulate houses at Marwell Zoo in Hampshire. Few are architecturally sophisticated. But all the buildings are highly functional and, from the animals' viewpoint, extremely hospitable environments to live in. They are not all built of reinforced concrete, steel and glass, they did not cost a small fortune to erect, and, as they become outdated, the zoos can replace them. Where, however, a zoo commits itself to a gigantic development involving a substantial capital investment, whether it turns out to be good or bad, the zoo is committed to keep it in operation for years to come.

Some errors in design are staggering, for they are so elementary. For instance, the Hagenbeck-inspired moated type of enclosure for exhibiting certain species like ungulates is without peer, but the translation of the concept is subject to dramatic distortion. We have talked to directors who reject the moated system. They say it is fairly expensive to install because it involves a retaining wall of considerable length and must have an effective drainage system irrespective of whether it is a dry or water-filled moat. Also, a large area of the enclosure is consumed by the moat itself. To us this seems a spurious argument, revealing an inability to plan and cost a development in the right direction. We have seen a project intended to provide a group of animals with a five-acre enclosure turn into a three-and-a-half-acre exhibit, with one-and-a-half acres of ditch. And, in one instance, a zoo in southern England produced a ditched enclosure that, when completed, had some of the neatest moats we have ever seen surrounding a piece of land the size of a back-garden! Nevertheless, moated enclosures are generally safer and aesthetically more pleasing than wire fences.

There are many types of barrier that can be used, depending on the species which has to be contained. Bars are still occasionally erected, although nowadays their forbidding appearance makes them unattractive and as a rule alternatives are found. Glass is becoming more widely used. It has proved an ideal material for primate cages, where animals which are susceptible to many human infections can be screened off from the public and, conversely, the public are protected from airborne infections that might be passed on to them. Reptiles are often kept behind glass, although a new technique of 'thermal caging' is being adopted. Instead of an actual physical barrier, freezer strips are placed across the front of the cage. Because the ambient temperature alone regulates their body temperature, the reptiles are unable to cross this icy strip. Birds can be kept in individual cages within a bird house by creating brightly lit enclosures along a straight public viewing area which is kept in darkness. The birds will not fly out into the dark and so are contained without wire or glass.

High tensile wire, chain-link, weldmesh, post and rails still have their place in the zoological collection and offer a wide variety of possible styles of exhibition when used with intelligence and in the context of the needs of the animals. The chronic suffering that can be caused by inferior caging is dealt with in the next chapter.

Zoos by definition should be academic institutions, yet they should also be exciting places to visit. The perfect collection is the one that fuses the science of wild animal husbandry and the art of wild animal exhibition. For instance, a good zoo will invariably have the foresight to enhance the appearance of the collection with carefully laid-out grounds, paying attention to small details like winding paths between shrubs and trees that are used to screen exhibits from each other. Screening not only improves the appearance of the grounds, it also adds the element of surprise and people enjoy the pleasant anticipation of wondering what they will see next. It is certainly a far better arrangement than entering the zoo's gates to face a field covered in cages, animal houses and random aviaries hotch-potched about the site.

Zoos only become boring places to visit when the animals are bored, when the enclosures and cages are poor and the general layout is uninspired. For many years the zoo establishment seemed unable to understand the need for artistry in wildlife exhibition. Some still seem unable, apparently believing the tiger in a concrete cell will more than satisfy a visitor's curiosity. Failing all else, they resort to amusements or funfairs. But a new concept did arrive, and its originator came from a circus animal-dealer background not unlike Hagenbeck's. The man was Jimmy Chipperfield and the idea was the safari park.

It is much to Chipperfield's credit that he alone saw and innovated a distinctly different type of wildlife collection from anything formerly seen in the United Kingdom, a concept that was highly successful, for the idea of driving in amongst wild animals instead of watching them from over a barrier captured the public's imagination to a degree few of his early critics could have possibly foreseen. It is just 12 years since the opening of the first park—Longleat—and since then the safari park business has grown into a multi-million pound empire, solely by selling the general public the impression that they are seeing animals at liberty. It is largely an illusion.

We are on record as being severe critics of safari parks and the reasons for this will be clarified shortly. Our basic criticism is that there seems to be an unfortunate lack of purpose behind many of these enterprises, save that which is purely lucrative. Furthermore, the standards of animal husbandry and the off-view accommodation is frequently poor. On the other hand, their potential for keeping and breeding many of the larger game species can hardly be matched by a conventional zoo; but so far we have not seen any evidence to suggest that that potential has been fully explored.

The great bone of contention for anyone concerned with keeping animals in captivity (especially if working in a zoo) is the constant insistence by the layman that animals in a safari park must be happier than their counterparts in a zoo, but there is no evidence to support this idea. In fact, apart from some special security problems, the safari park requirements for animal housing, veterinary facilities, diets and so on are the same as those for similar species in a conventional zoo. If the public were to see behind the scenes they would probably quickly come to the conclusion that in many instances the zoo animal was decidedly better off!

A popular feature of several parks is the boat safari, a short cruise on the lake during the course of which the visitor can feed the sea-lions and see the chimpanzees on their island. The consensus of opinion amongst the public seems to be that this is a delightful way to keep chimps, free to do as they like on their own island, something far preferable to the standard type of zoo cage. But is it? In one park we visited the island was a man-made pile of earth a few yards off the main shoreline, which, apart from a covering of grass, had little that could possibly provide any interesting day-time diversion for apes of this high intelligence. The sleeping quarters, buried discreetly beneath the soil, were built from large concrete sewer pipes that led into two small barred cages. As there was no power leading to the island, all heating and light was provided by calor gas. In short, the accommodation was very crude. A chimpanzee in conventional zoo accommodation which might only be classed as reasonable would have heated sleeping quarters for the night and some form of occupational therapy. After all, even bars can provide a climbing frame.

Sewer pipes seem to be a popular building material in these places. Another park we know had a similar style of sleeping accommodation for its baboons. These were bunkers buried under the grass of the enclosure: thus they neither blighted the landscape nor offended the public. They were cold, dank and dark. In severe winter weather they became flooded with water, through which the unfortunate baboons had to wade if they wished to seek shelter on one of the wooden ledges within. In the same enclosure there were a number of eland antelope and their accommodation was little better. It was basically a three-sided wooden shelter with hay-racks fixed upon the back wall, but, as with the baboon bunkers, its floor was below ground-level and as no form of drainage had been installed, it too flooded. We well remember watching these animals standing forlornly in the shelter, unable to lie down because their straw bedding was floating on water. These tragedies are not due to inadequate finance, for the safari park is an incredible financial success. They are born of inexperience, or possibly lack of concern.

More weaknesses can be exposed by further examples. Take the case of the hippopotamus. At one park the hippopotamus was kept in the lake: a charming spectacle, yet technically impracticable. Under conventional conditions this species is provided with a pool which can be cleaned regularly and, most important, heated. A lake dispenses with the need for cleaning as faecal matter is broken down organically, but it does not dispense with the need for heating. We have been told of a safari park that lost a hippo which had run out on to frozen water and fallen through. Unable to surface and doubtless overcome by the freezing cold, the animal died almost immediately. But in a later instance the problem was not one of heating water, it was one of security. Alongside the lake ran an enclosure surrounded by a six-foot, chain-link fence which was just not strong enough to contain a hippopotamus, so periodically the animal crashed through it and ambled off into the English countryside.

These escapes had their amusing side. On one occasion the animal went for a stroll down the village high street in the early hours of the morning, but when stopped by the park staff he simply trotted back to his enclosure with an air of indignation that seemed to imply that they were rather mean spoilsports. He was definitely an amiable creature but obviously the situation could not be allowed to continue. In time he would doubtless encounter someone who would over-react and a severe accident would result. So the park authorities decided to move him from the lake to a more secure location. But they had no suitable accommodation in which to put him at night. Their solution was to purchase an old railway wagon, the type with double doors along each side, and the poor animal was confined within this for over a month while an adequate ditch was dug around it. Without room for movement, without proper ventilation, the hippo sweated and grew distraught, periodically buckets of water were thrown over him to keep his skin moist but this only eased the suffering slightly; otherwise the torment must have seemed interminable. When the ditch was completed the railway wagon was given a rustic appearance with a covering of pine bark. Since when, none of the thousands of visitors who pass it have raised a single voice in protest at the creature's conditions of confinement.

Bears have also been kept in railway wagons in safari parks, as have lions. We could continue with examples of the crudity of much of the housing in safari parks, like the rhino house that was so small that when the rhinos were bedded down for the night it was impossible for them to turn round—if they went in nose to tail, they remained nose to tail!

As far as the general public are concerned, what the eye does not see, the heart will not grieve over. This is effectively the key to the financial success of many unworthy establishments: people are not critical enough; they willingly believe what they think they see. Nevertheless, it is amazing that the public are so slow to react to the more dramatic reports of escapes, something that should not happen if the wildlife collection is properly managed. In one year in a single safari park, numerous species escaped.

A whole herd of eland antelope smashed through the fence of their retaining yard and were found grazing near the main entrance. All were recovered without injury.

Baboons escaped with monotonous regularity and were usually recovered without undue drama. However, on occasions, some had to be shot. (For instance, when one new park was opened, a number of baboons escaped, some even getting on to a nearby motorway. The hunt and the subsequent shooting were well-documented in the national press.) More disturbing, however, were the escapes within another park, for although the baboons were easily returned to their enclosure, they habitually raided the park restaurant, covering the outside tables and serving hatches with an abundance of faeces and urine. The health hazard to the public is quite terrifying, as can be vouched for by anyone acquainted with the diseases of primates that are directly transmissible to man.

Sea-lions made constant escapes into nearby lakes where they ruined the fishing.

Several crown cranes got out, although not all were recaptured. Those that remained loose would have presumably died when winter set in and food grew scarce.

Other escapees included rhinoceroses, elephants, giraffes and ostriches. In each instance the animals were secured fairly rapidly and no harm was caused to them, the staff or members of the public. But, as a chronicle of escapes covering only one year in one safari park, it must be considered a serious indictment against the park's professional standards.

During this period the most dramatic escape was that of a lioness which got out of the lion enclosure some time during the early hours of the morning on a Bank Holiday weekend. After several abortive attempts to recapture her, the park authorities eventually had no alternative but to issue the order to shoot her in order to safeguard the public who, by this time, were starting to arrive by the hundreds. The incident was given considerable coverage on television and in the national newspapers, and the park management was obliged to give an explanation for the occurrence. Their version was that the lioness had been harassed by her companions, so in a desperate search for peace she had broken through the perimeter fence and run away from the source of torment. The implication that exceptional pressure had caused an exceptional reaction was fair enough, but the management failed to admit that the twelve-foot fence through which she had escaped did not go beneath ground level along its entire length. For some time prior to the escape several staff had commented on the fact that rabbits freely passed beneath the fence once they had excavated a small depression beneath it, and several enclosures were classified as unsafe. No one listened of course, and it was not until the lioness had escaped and been shot that remedial action was taken. Today, if someone cared to dig down below the fence-line they would find that a further piece of chain-link has been woven on to the original, a clear case of shutting the stable door after the horse has bolted.

Chapter 6

# Behaviour in a straitjacket

Do animals in captivity pace up and down because they want to get out? Why do some species, like elephants and bears, stand on the same spot and sway from side to side? And why do some birds pull out all their feathers and mammals groom themselves until their bodies are covered in bald patches? Is it normal for animals to chew their tails until only a vestigial stump remains, or eat their faeces?

That this type of behaviour exists is undeniable and most of us have probably seen this sort of thing in zoos or wildlife parks at one time or another. But is this behaviour normal? Can it be seen in the wild? If not, then why does it happen and can it be avoided?

Animal behaviour is understandably a complex subject, displayed as it is in a wide variety of ways by a multitude of species, so a detailed discussion is beyond the scope of this book. But in general there are specific patterns of behaviour for specific situations, like courtship and mating, territory, feeding, threat and appeasement, amicable behaviour and so on. The subject is further complicated by the acknowledged existence of distinctions between behaviour that is instinctive and that which is learned, but we shall deal with these factors, where relevant, in more detail later in the chapter. Quite simply, every living creature, from the lowliest invertebrate to an animal on the highest rung of the evolutionary ladder, is equipped with sense organs and behaviour patterns that are essential for survival in a hostile world. They are born with these senses and retain them throughout their lives, irrespective of whether those lives are spent in the natural habitat or in captivity. And here is the rub. In order to care properly for any animal in captive conditions, it is imperative to recognize that the composite creature is more than just flesh, blood and bone—it also has natural desires and feelings.

Obvious as this fact is, relatively few zoo owners pay enough attention to this aspect of animal welfare, and as a result many captive specimens have both senses and behavioural responses placed in an environmental straitjacket. When emotions are blocked off in this way they don't cease to exist but, like waters held by a dam, they build up until the pressure becomes so great that they burst forth in torrents of frenzied abnormality and neurosis. A deluge evidenced by the characteristic symptoms of pacing, swaying and self-mutilation, accompanied by the less dramatic, yet equally regrettable, rivulets of apathy so often seen in below-average collections in Britain. Let us look at an elementary case involving the misinterpretation of an animal's biological needs.

The European badger (*Meles meles*) is not an uncommon inhabitant of zoos and wildlife parks, which is hardly surprising for they are as attractive as many of the tropical species and absolutely enchanting to watch. Also, prior to recent protective legislation, they were obtained fairly cheaply, a primary incentive for a lot of prospective zoo purchasers.

In the wild the badger inhabits forest, woodland, dense shrub, in fact most areas that provide plenty of cover, although it usually likes to build its sett in an open sunny glade, and preferably in dry, sandy soil. The sett, or den, is a complicated affair, consisting of numerous passageways that serve as airholes, entrances and exits, plus several sleeping chambers. The badgers complete their home with refinements like a scratching tree above ground, and, when young are being reared, an excretion site. Setts may be used for decades and in certain instances there is evidence to suggest that the same one has been used by successive generations for centuries. Although essentially nocturnal animals, badgers, as any keen amateur naturalist will know, are frequently observed above ground during the day-time, either roaming around in search of food, or simply sunning themselves. Their diet is varied to say the least. Classified as carnivores, which indeed they are, it is surprising to learn that three-quarters of their food consists of vegetable matter in the form of seeds, fruits, mushrooms and roots, whilst the remainder is made up of any animal, alive or dead, that the badger can cope with. When they mate, they pair for life.

From this brief outline of the badger's life-style, one would logically assume that in order to make a good exhibit with this species it would be essential to provide them with an underground den and a sizeable area of ground for digging, exploring, sunbathing and playing. A reasonable amount of cover would be needed so that the animals could retire from the presence of spectators, but at the same time they should remain visible to the public without having to endure any stress. Such an ideal style of exhibit can be seen at the Norfolk Wildlife Park, Great Witchingham. Here it is possible to see the badgers through the glass windows of their underground sett without the risk of disturbing them, although they will often appear on sunny afternoons in their outside enclosure, where one can watch them being fed.

It's an idyllic little enclosure set amidst a fine collection of European animals, yet, without wishing to denigrate the efforts of Norfolk, it is only produced of basic common-sense. The badgers want peace and security, the public want to see the badgers, so a solution is found for both interests at Norfolk—to the detriment of neither.

But do all zoos consider the customers and the inhabitants so sympathetically? Frankly the answer is no. Of the badgers seen in British collections in recent years, two enclosures have left a lasting impression upon the authors. Contrasting so dreadfully to the Norfolk exhibit, we feel they must be mentioned.

In many ways the two were not dissimilar constructions, one cage being rectangular with three granite walls and a wire front, the other an oval wire enclosure. Both had concrete floors, and both were provided with small wooden boxes for the badgers to sleep in, but, apart from a water bowl, they were devoid of any other articles either to occupy the animals or to enhance the general appearance of the exhibits. The errors are blatantly obvious: badgers dig, but a concrete floor immediately thwarts attempts to fulfil this normal activity. And as they are essentially nocturnal, secretive creatures which like plenty of cover to hide in, it is unlikely that they would willingly venture into an exposed area with onlookers only a few feet away. Nor did they. In both these collections the badgers spent every minute of daylight hidden inside their boxes, which incidentally were cramped, stuffy and in every way inferior substitutes for the highly refined setts dug by these animals in the wild. Looked at from the purely commercial viewpoint, such a situation was unsatisfactory because the public never saw the animals. Not only would the badgers have been much better off left in the wild, but their captivity served no useful purpose.

Eventually one of the collections received so many complaints from visitors about this and other similar exhibits in which the animals could not be seen that the management decreed that those species which hid from the public would be shut out of their sleeping quarters during visiting hours. This expedient, though hardly intelligent decision, resulted in the two unfortunate badgers rushing blindly about the enclosure from ten o'clock in the morning until the closing of the park at night, unable to rest because their instinct told them to flee, yet there was nowhere to flee to.

This example highlights some very common failings found in zoos and wildlife parks. Collectively they might be simply qualified as failure to acknowledge the existence of the normal requirements of the species exhibited, but it is possible to be far more specific.

Locking animals out of their sleeping quarters to ensure that the public gets its money's worth is a widespread practice that accounts for a considerable amount of unnecessary stress and suffering in zoos. It is essentially an admission that the accommodation provided is badly designed because the public should be able to view the animals without such drastic measures being taken. Naturally the degree of stress will vary according to the species concerned, and the size and quality of its enclosure.

Safari parks, for example, turn nearly all their animals out of the sleeping quarters in the morning, not allowing them back until the park closes for the night. But in this circumstance it is not necessarily a bad thing. Most of these drive-through parks have large sections that contain a fair amount of trees and shade, so the animals are able to move away from the noise and smell of excessive traffic and find places to rest peacefully. Certainly stressful pacing behaviour is a rare sight in the paddocks in these parks, although it has been seen in the indoor accommodation which is often confined. Unfortunately, most safari parks spoil their reputation on this point of relative freedom by insisting that the animals are kept within certain areas that facilitate public viewing. This means regular round-ups (and therefore a new type of stress) of giraffes, zebras, rhinoceroses, lions and tigers by rangers on tractors or in Landrovers, which in turn means a great reduction in the size of enclosure. On one Bank Holiday weekend in a safari park the authors watched a group of 17 giraffes, which was effectively kept on an area measuring approximately 150 yards by 50 yards for most of this time. Yet this was a busy period in which they needed greater freedom of movement in order to get away from the crowds, not less. For the public who drove past this herd during those four days it must have been an impressive sight: so many giraffe in acres of land. But they might have been very surprised had they known how little of this acreage was being utilized. This type of practice is an inhumane and stupid error on behalf of the safari park operator, for the only single advantage that the safari park has over the conventional zoo is the size of the enclosures; yet this advantage is constantly abused.

But the traditional zoo displays by far the greatest number of examples of nervous behaviour caused by exposing timid, shy creatures to a constant stream of visitors, just by denying them areas of seclusion and security. One zoo we visited on a summer's day in 1974 had every species exhibited shut out of their shelters and sleeping boxes. Those species that would normally flee from danger rushed madly about their cages, and those that would normally hide, pathetically tried to conceal themselves by 'hiding', absolutely motionless, in the cage corners.

At another zoo we saw a very nervous trio of Barbary sheep

dashing about a tiny enclosure and leaping against the walls of their house, simply because the zoo's management had insisted that the animals should be shut outside at peak periods so that the public would be sure to see them. Had someone merely opened the doors of the sleeping quarters, in most cases this behaviour would have ceased immediately. Yet the key factor giving rise to this state of chronic tension is often the most ignored. For it is not just that the animals are shut outside in an exposed area, but that the actual presence of people around a confined space infringes on their natural 'flight distance'. The two factors often seem incompatible, for the time that the animal most needs some form of peace and seclusion is the time when it is most in demand—peak viewing periods.

Flight distance is the limit between the animal and an enemy, predator or some other object that seems to threaten it. Provided that 'threat' remains beyond this limit, the animal feels safe and can pursue its normal activities quite happily, but once this invisible boundary is crossed it must take flight and run, hide, dive into its burrow, as its instincts dictate. The distance varies according to species and within species according to the individual and, as a captive animal becomes tamer, its flight reactions will usually tone down considerably. Nevertheless, toned down or not, flight distances exist and should always be taken into consideration by zoo management.

When flight distance is related to captive conditions, one can find that a cage, which at first sight seemed reasonably spacious, suddenly changes into a frighteningly restricted box. Imagine an animal with a flight distance of 15 feet in a cage 30 feet square. If a line of people stand at the front of the cage, the actual area that remains 'safe' ground for the inmate is 30 feet by 15, but if these people then stand on all four sides, the cage is further reduced and the animal is actually forced to move to a spot exactly in the centre.

Certain zoo owners will probably insist that this is a gross over-simplification of the facts, and admittedly this is true. But when the theory is replaced by a practical demonstration, the results can be extremely horrible and distressing to see. We had the misfortune, over a period of several months, to observe the following case of hypertension that was caused primarily by close confinement.

A wildlife park kept seven grey wolves in two small pens, one measuring approximately 8 metres by 3 metres, the other—an almost circular pen—about 4 metres in diameter. The two pens were joined by a wire tunnel barely half a metre high so, as the wolves crossed from one pen to the other, their backs rubbed against its roof.

The exhibit's history reveals much of its failings. Originally a pair of wolves were purchased by the park owner (a man who openly admitted that he knew nothing about wild animals) and a makeshift pen was built to contain them. They bred. Finding the pups such a valuable attraction for the public, the owner decided to keep them, hence the group of seven animals and the building of the second, circular pen and communicating tunnel. By any standards the enclosures were crude: chain-link stretched between poles to form a retaining fence; rough concrete floors, stupidly laid so as to encircle a number of trees which ultimately died; and one wooden shelter into which seven wolves could just manage to squeeze.

In time these unfortunate creatures fell victim to two distinct forms of torment, the previously mentioned infringement of flight distance and, in this restricted environment, the pressures of overcrowding. Squabbles broke out amongst the group far more frequently than is normal with the species, aggression was excessive, and the chain-link was repaired in dozens of places, clearly indicating where the frustrated animals had torn at it in an effort to escape. But it was on busy days that tension reached its peak. Three-quarters of the exhibit faced the public, so in order to seek a measure of security the wolves were forced to sit as far back as the rear fence would allow, or try to hide in the open-fronted shelter. Often two or three wolves would take possession of this and attempt to keep the others out by snapping at them as they approached. But, sooner or later, another would manage to dive in, forcing one of the original occupants to move out again.

So, as the park grew more crowded, the tension in the wolf pack gradually escalated and a vicious circle became set. As the public encircled the pens, the wolves would rush back and forth through the tunnel, into their box, then out, back through the tunnel, in a constant state of panic. On one particularly busy Sunday, an especially unruly group of visitors so terrified the animals with their shouting and gesticulating that the lead male leapt at the fence (2 metres high plus an overhang) and almost succeeded in scrambling over. Indeed, had it not been for the quick reactions of a nearby keeper who pushed it back with a broom, it might well have done so. Wolves are often kept in enclosures with fences of this height and sometimes less, yet they do not attempt to jump it. But in this instance panic was the spur. The affair had a sad, although in our opinion, fairly typical ending. Realizing that he had a severe problem on his hands—that it was only a matter of time before the wolves escaped by either tearing a hole in the wire or by leaping the fence, the park owner was coerced into building a spacious new enclosure. A frugal man by nature, he agreed to purchase building materials but, regrettably, could not be persuaded to employ men to carry out the construction work, insisting that the keepers could easily undertake this task in addition to their daily duties. Not surprisingly, by autumn of that year the new enclosure was still far from completion, yet the risk of escaped wolves running amok in the nearby village grew more likely each day. So the owner loaded his rifle, went down to the wolf enclosure and shot four of them through the head. He justified his action on the grounds of public safety, although he was not slow to remark that it would also make a saving on winter food bills, and started the off-season economies by having the wolf cadavers fed to the lions.

Could this park owner have known of the problems that would arise when he purchased that first pair of wolves? It would appear that he did not, but we do not believe that that constitutes anything like a valid excuse for what happened as a result of his actions. Information on wolves is well documented and they are known to do well in captivity so long as they are not too confined. It would have taken very little investigation to establish this fact. Also there are a number of fine wolf enclosures to be seen in British zoos, including the superb Wolf Wood at Whipsnade where the wolves are shown in twin enclosures that are thickly planted with pine trees. John Aspinall's private zoo, Howletts, in Kent, provides another fine example. Here a large woodland area has been fenced off for the wolves, and Mr Aspinall has established a strong relationship with his groups and often goes into the enclosure to play with them. The latest, and probably one of the most exciting so far, is a new Wolf Wood at Edinburgh Zoo, where five wolves have been established in an extensive area planted with birch trees. None of the specimens exhibited in these collections display tense, nervous behaviour of the kind we have been discussing, and this is simply because they have been provided with a home and not a prison! Not only must the captive environment be compatible with the specimen's flight distance, but it is equally essential that the staff are familiar with it, as they should be with all aspects of animal behaviour. But zoo staff, with no training at all, are often employed by zoos and safari parks and are expected to cope with a wide range of behavioural problems. The results can be catastrophic.

On one occasion in a safari park the writer watched two newly recruited rangers open the gates to the retaining yard of the giraffe house in order to get the herd in for the night. It is a reasonably easy task, for the giraffes, anticipating their evening feed, readily come in from the main paddock and are often found waiting at the gates if the lock-up is later than usual. As the rangers closed the gates behind the herd they suddenly realized that one female was still outside. She appeared almost immediately, walking gracefully across the pasture in the direction of the yard until, some distance from the gates, she stopped. One man unfastened the gate, stepped aside and waited; she remained motionless.

The rangers waited patiently, one on either side of the gate, occasionally calling softly to her and holding out handfuls of hay and leaves. From time to time she moved forward, faltered and then stepped back. What the men failed to realize was that the giraffe desperately wanted to enter the yard but that it was they who were stopping her by standing on either side of the entrance. To pass between them, the giraffe would have had to overcome its flight reactions in two directions at the same time, a psychologically impossible feat!

As darkness began to fall the herd in the yard, still waiting to be admitted to the giraffe house and food, grew restless, the lone female became increasingly agitated and the two men, thwarted in their efforts to get home, became bad-tempered. This only made matters worse and the stalemate finally exploded with the herd bolting from the yard, back into the main reserve to join the isolated female. In the ensuing panic the young rangers were nearly trampled and a number of giraffes grazed flanks and legs against the gates and metal fence-posts.

All this confusion was caused by ignorance of a basic principle that countrymen know almost instinctively. A man herding cows or sheep will step back well away from a gate to allow the animals to go through. He knows very well that, if he stood by the gate-post, the animals would be afraid to enter because to do so would place the man within their flight distance.

Quite apart from the stress caused to the animals, a situation can arise when the keeper, albeit unknowingly, is exposed to considerable danger himself. With most species, the more one intrudes into the flight range the more intense becomes their behaviour, changing from mild fear by degrees to intense fear and eventually breaking-point. The lead wolf reached this critical stage and leapt a fence far higher than it would normally attempt. Some species, if crowded into a corner, will suddenly attack the object that threatens them. This well-known principle is used by the lion trainer in his circus act. He moves towards the lions so as to gradually raise the level of aggression, which, much to the audience's delight, causes the lions to snarl and swipe at his whip; when he backs off they quieten down again.

It is quite definitely not a game for novices. Yet we know of an incident in a safari park when one of the senior rangers attempted to chase a tiger into its sleeping quarters armed with only a stick. He was fortunate to escape with minor lacerations on his arms. Incidents like this are a positive indication of lack of initial training.

Restriction of space, lack of seclusion and pressures on flight reactions are not the only changes the animal can experience when put into captivity. Removal of the time-consuming labour of finding food, added to confinement in an impoverished environment, reduce the natural stimuli and often prevent the fulfilment of natural instincts.

Niko Tinbergen, the Dutch ethologist, believes that animals kept in captivity often show behaviour that, by comparison with their behavioural activities in the wild, makes no sense. In fact it misfires. He compares living to a tightrope act: there are many ways to fall but only the one narrow cord leads to success. As captivity always represents an artificial environment to which some degree of adaptation is necessary, the risk of falling is increased rather than reduced.

Some of the safari and wildlife parks are certainly guilty of ignoring the fact that captivity can only be artificial, even exploiting the opposite view by encouraging the public to see animals in natural surroundings. But how can a tiger in an English meadow be in natural surroundings? Far better that the captive environment be based on a sound assessment of the animal's needs rather than a cheap mimicry of its natural habitat, for the former takes the animal's welfare into account whereas the latter simply considers the spectacle displayed to the public.

For instance, earlier in this chapter we mentioned that by keeping badgers on concrete they are deprived of the normal and instinctive pursuit of digging; on the other hand, if a suitable piece of ground were fenced off, the badgers would soon dig out. The solution can be produced in various ways. Wire may be buried under the soil, or the wall of the enclosure can be built down several feet below ground level and an inward-facing lip added to prevent escapes. Even a long-established concrete cell could be radically altered by the addition of a couple of barrow-loads of soil, for the badgers would spend hours ecstatically scratching about in search of beetles, earthworms, grubs and roots. Although this would mean a considerable amount of clearing up for the staff each day, wouldn't it be worth the extra effort?

Essentially the maxim for all wild animal collections should be that if the animal flies then give it space to do so; if it swims then provide it with a pool; if it climbs let it have suitable climbing apparatus; if it digs give it the opportunity to dig. It is pointless to subvert a natural instinct and criminal to subvert it to the extent of denying the performance of a normal behaviour pattern, as was the case with the badgers. But all too often this is exactly what happens.

The wildlife park containing the cramped wolf enclosure was full of similar behavioural prisons. A pair of wild boar were kept in a rectangular pen which was built of concrete blocks surmounted by a chain-link fence. The walls of this pen were too high, so unless the animals rested their forefeet upon them to peer and sniff about, they were totally cut off from the sights, sounds and smells of the outside world. As the floor was of rough cement another possible diversion was denied the poor creatures, for boars are natural bulldozers and like to snuffle about in the soil and in the wild spend most of the time searching for roots and plants. To aggravate the situation still further, their solitary sleeping den (built into an adjoining outbuilding) was far too small, and the boars often fought for possession of it. Usually this meant that the first one in was able to obstruct the small entrance by threatening its partner with its formidable tusks, so it was not uncommon to find one of them sleeping in the corner of the outside pen. Cleaning the boars' pen was a major problem because to do this it was necessary to shut them inside, not only a difficult thing to do but, if achieved, fighting broke out almost immediately, so, understandably, the pen was cleaned infrequently.

Incredible as it may seem, in view of the conditions described, each year the boars had young, but owing to the constant stress factor the sow invariably devoured them within a few days of birth.

Nearby a trio of Humboldt's penguins lived in an enclosure with a pool only marginally bigger than a household bath. And a lone masked palm civet, an essentially arboreal species, was confined to a cage little more than a metre high. Without doubt, the most terrible example of all was that of a fruit bat housed in a plywood box only slightly larger than a tea-chest. The box had been laid on its side, a sheet of glass fitted across the open end and a square of wire-mesh tacked on to the ceiling for the bat to hang from.

Of the nearly 2,000 known forms of bats, the vast majority make unsuitable subjects for the zoological garden, although in time, when our state of knowledge improves, doubtless more species will be exhibited. But at present the only notable exceptions to the rule are the Old World fruit bats, and they often do exceedingly well in captivity. London Zoo have kept no less than 14 forms of fruit bat, a considerable number of which have given birth, and one specimen lived there for over 17 years.

The bat in the wildlife park was an Indian form, commonly known as a flying fox, so-called because of the fox-like appearance of its head. In the natural state they roost in large colonies in trees, caves or on cliff ledges, flying off at sundown to their feeding grounds, which are often as far as 15 or 20 miles away. Equipped with acute senses of smell, hearing and vision, it seems that sound orientation plays only a minor part in their lives and that they depend almost entirely on their keen eyesight, although some of the smaller species of fruit bat have been found to possess echo location apparatus. Yet here this fruit bat was shut in a poorly ventilated box, unable to see, or smell or hear anything that existed beyond the glass-fronted barrier. Unable to utter a sound that wasn't absorbed by the plywood walls, unable to fly or even spread its wing membranes to their full extent.

Despite the deprivations, all these animals survived, the bat and the palm civet were sold to an animal-dealer and, at the time of our last visit, the wild boar and penguins remained as they were. Here the lack of four normal activities, swimming, climbing, digging and flying, plus the isolation of senses like sight, smell and hearing, are found in four exhibits in just one substandard wildlife park.

Some species—because of their physiology and in particular their psychological make-up—once settled into an artificial environment, seem hardly to notice the fence surrounding them. This is especially true of those creatures that spend the greater part of the day feeding, like the vegetarian browsers and grazers. Of these, the animals with simple stomachs like the zebra and the horse eat slowly and continuously all day. While those with more complex digestive systems, the ruminants, eat rather more quickly for shorter periods and then lie down to chew the cud.

Ruminating is a method of eating developed by animals to escape from their enemies. They can quickly take into their bodies large amounts of vegetable matter and then move from an exposed area to a place of security to relax and enjoy their food. Chewing the cud is a process where the animal brings up boluses of partly digested food and chews it more thoroughly before swallowing it again. The members of the deer family are ruminants, as are giraffes, pronghorns and all the *Bovidae*, a vast group containing animals like bison, antelope, hartebeest, wildebeeste, oryx, musk ox and ibex.

As a general rule all of these species, as long as they are maintained in reasonably spacious areas of grassland and kept in a social grouping within the normal range, will remain placid and contented in captivity because they are constantly occupied with eating. Conventional zoos rarely have enough space available to accommodate these types properly (although there are exceptions), whereas the safari park and the wildlife park often do have large areas of pasture suitable for maintaining and displaying them. It would therefore seem a natural progression for these places to concentrate on the herbivorous species, leaving the more demanding subjects for specialist collections. We have, however, noted failings in the organization of these exhibits, the most commonplace being the gathering together of abnormal social groups. Zebras feature significantly amongst these observations.

During the rainy season in Africa the common plains' zebras often form herds of 12,000 or more, but after a short time they disperse, forming much smaller groups. These may be family groups comprising a stallion and one or more mares and their foals, or pure stallion groups, but the sexes are never equally represented. A stallion is in fact very possessive of his mares and will quickly see off any strange males which dare to approach them. Zebras can be extremely aggressive animals and although fights between stallions are rarely serious in the wild state, in the restriction of captivity they can easily lead to severe injuries, and even death.

Owing to a surfeit of stallions in captivity they are understandably cheaper to purchase than mares and much easier to obtain. Thus the example of the stallion herd so often found in the wild has been adopted by a number of zoo owners in order to create an exhibit. Keeping only males seems a pointless exercise but, in view of the fact that stallion groups have been known to remain together for many years quite naturally, there seems little reason to suppose that the practice causes stress or suffering. But add a female to a stallion group, or put together a number of mares with three or four stallions and an explosion is guaranteed. Not only does this happen far too frequently, but one zoo group moves animals like zebras from site to site and establishes new groups with such total disregard for normal sex ratios, it is almost unbelievable.

The results of this practice speak for themselves. We know of a collection, to which a small herd of Grant's zebras were sent —four stallions and two mares. At night they were housed separately in stables adjoining the giraffe house, so the opportunity to fight did not arise. On the other hand, during the day when the group occupied a large area along with eland antelopes, rhinoceroses and giraffes, squabbles broke out frequently, so the rangers on the section were told to keep a careful watch on them. Driving a Landrover between battling zebras resolves the situation quite dramatically, but not permanently. In due course one stallion had to have a two-foot gash on its rump stitched up, another had its genitals so savagely bitten that the scrotum had to be sewn together. Thankfully both animals recovered.

The group was finally dispersed amongst other parks so we know nothing of their ultimate fate, save one specimen. This animal, one of the stallions, was anaesthetized with a dart-gun and moved to a small paddock by a lake, where he was then injected with an antidote and recovered enough to get to his feet, although understandably he remained rather glassy-eyed and unsteady. It is normal procedure under these circumstances to put such specimens in a quiet, secluded place devoid of obstacles so that they can recover in peace and without any risk of injuring themselves. In this case however, the zebra was released directly into the paddock before he had fully regained his senses, and no one remained to watch over him. Such stupidity resulted in the semi-comatose stallion falling headlong into the lake and drowning. Strictly speaking, this is not a behavioural example, rather it is the sad conclusion of a situation created by ill-advised grouping and the subsequent attempts to rectify it being carried out by incompetent staff.

Such situations should never arise of course, and, as we have already said, many of these continual feeders will settle down to life in captivity quite easily as long as they are kept in social groups comparable with those found in the wild and are provided with spacious, grassy* enclosures. Whipsnade was a pioneer in exhibiting herd animals in these conditions and, despite a shaky period in its evolution when the collection became rather run down, it remains one of the best places to see them. Of the newer developments in our zoos and wildlife parks undoubtedly the most important by far, particularly with reference to these species, is the Marwell Zoological Park, near Winchester.

Marwell is neither a drive-through park nor a conventional zoo. In general the enclosures are large and one often has to look

* There are problems associated with keeping animals on grass which can be overcome if adequate veterinary precautions are taken: these are referred to elsewhere.

hard in order to identify all the species contained in them, but this is in no way disadvantageous and can often prove a pleasing and rewarding experience. Among the animals which enjoy the relative freedom of this excellent park are many species rarely seen in this country, including scimitar horned oryx (one of the largest breeding groups in captivity), sable antelope, European bison, Przewalski's horse, wild ass, Père David's deer, and nyala. Giraffe and a variety of deer can also be seen, plus three distinct races of zebra and a superb collection of cats.

The graceful little impalas are probably proof positive that Marwell has achieved a unique harmony between animal husbandry and captive environment. Although not an uncommon species on the plains of Africa, impalas are rarely seen in zoos because their acute senses, agility and speed make them very difficult to maintain. Disturbances that would merely upset the average animal will spur this slender and elegant antelope into a sudden burst of running and leaping. They can jump about ten feet from a standing position! In a wire enclosure this can easily lead to broken legs and necks and it is certain that only a minority of British collections would possess the skill required to keep these creatures alive for any length of time.

Other species may be more aware of their captivity than the herbivores, but will adapt well to it provided they are also kept in near-normal social groups and the space provided is not too small and uninteresting. Specialist feeders, such as the big cats, are content in the wild to sate their appetite with one large meal and then spend much of their time sleeping it off. One might say that they tend to be lazy and rather aloof from their environment. As long as they can retreat out of sight of the public if they wish, and are provided with modest requirements like a log to scratch, a resting bench, or—if an arboreal species such as a leopard—branches to climb, they will adapt well to captivity.

Unfortunately most zoos keep these species in small cages that are totally devoid of interest and in this type of environment they can be seen pacing up and down for hours on end. Examples are almost too numerous to mention. In recent years we have seen most of the big cats—tiger, lion, leopard, jaguar and puma—exhibited in this way.

Even such a well-regarded zoo as London had, until early in 1976, a cat range hardly worthy of its reputation. Built 100 years ago, it was, in its time, the largest and considered by some to be the best cat house in the world, but ideas on animal hygiene and comfort were very different in 1876 from those held today. Therefore, in recent years the old cat range caused considerable criticism on the grounds that it was aesthetically unpleasing from the public's viewpoint, and environmentally unsuited to the animals it housed.

The new cat enclosures, opened at the time of London's 150th anniversary, are pleasantly landscaped areas designed to present a pleasing panorama to the visitor and an environment that is interesting for the cats to live in. It is arguably the best of London's redevelopment schemes.

Yet in an age when many of the older Society collections are doing their best to modernize their exhibits, a lot of commercial zoos are simply mimicking the type of caging developed 30 or 40 years ago. Thick metal bars are not often used because they have become generally unacceptable to the public and most of these cages are built from timber and chain-link, but in principle they conform to the conditions created in the square or rectangular barred cages of earlier this century. Under these conditions lions, tigers and leopards will often develop a form of pacing that is stereotyped rather than normal. (Pacing can be a normal activity even in an ideal environment, either as a form of exercise, or, most commonly, shortly before feeding-time when the cats become excited at the prospect of their food and will dash up and down the fence-line, eyes focused on the spot from whence they expect the keeper to appear.) Stereotyped pacing is not associated with either of these; it is a form of behaviour that occupies the animal for most of its periods of activity without any apparent normal stimulus bringing it into operation.

A Siberian tiger we saw in a small south coast zoo in 1975, kept in a cage of this type, displayed the characteristic symptoms of this distorted behaviour. It paced along the wire until reaching the corner where it turned, still facing the wire, and retraced its steps to the other corner. It moved back and forth for hours, never turning inwards but always looking beyond its enclosure in a state of unrelaxed tension.

Stereotyped movements in wild animals are practically never due to a single factor. If a generalization had to be made it would be that the root cause of the abnormality is the blocking-off of emotional outlets—or the shutting up of their many mysterious senses and behaviour patterns.

Nearly every species will show some aspects of this type of behaviour if it is kept in an impoverished environment, lacks occupation or is unable to fulfil a normal instinct. It could easily be thought that weaving from side to side and pacing up and down are normal habits, so routinely are they found in zoos. In fact our own observations lead us to the conclusion that at least one such case can be found in seven out of every ten collections in the British Isles. This figure is merely an estimate based on observations made by the authors over a number of years. Obviously some of these collections contain relatively few deprived specimens, whereas others have a preponderance of animals that are clearly psychologically disturbed.

Stereotyped movements are a sure indication of wrong treatment. They should never develop, but if they do it is possible to rectify them, providing action is taken quickly, for if these abnormal patterns become firmly established the effects are often irreversible. One extremely sad example that springs to mind is that of a polar bear that was purchased by a small zoo and kept in a tiny, dark cage for many years. In the squalid, cramped conditions of this cage the bear's movements were terribly restricted, so badly restricted that it was only able to exercise by walking diagonally across its cell. Every day it would pace across the cage until its nose touched the bars then, unable to turn around comfortably, it would shuffle backwards until its posterior bumped against the concrete wall that formed the rear of its prison. No sensible purpose was served by the behaviour, no reason was applied to the action, the bear's emotional outlets had simply been shut off by the constraints of its environment. In time the pattern became firmly established and the bear's rump and head, rubbed raw and scabbed by its monotonous actions, gave grotesque witness to the creature's plight. Here we have a zoo, apparently lacking finance and experience, keeping an animal, which needs large, specialized quarters, in a dungeon. Polar bears are acknowledged big attractions in any zoo and this was supposedly the justification for the exhibit. Eventually with the appointment of a new director a number of redevelopment schemes were started, including a fine polar bear exhibit, into which this unfortunate animal was moved and provided with a swimming-pool and a mate. In due course the bald patches on the rump were covered with a luxuriant growth of hair and the scars on its muzzle faded, yet if such visible signs told of improvement, they told a bitter lie. The deepest scars borne by the animal were not on its body, they seared through its mind: for as a result of its long imprisonment, the bear had become stereotyped. The sad white giant still plodded forwards to the edge of the pool, then shuffled slowly backwards to the point where its cell wall should have been; only one-fifth of the large exhibit ever being used by it. Whatever emotions remained were slumbering, never to be re-awakened.

Bears feature significantly in recorded cases of stereotyping, probably because for centuries it was customary to keep them in deep, circular pits, where—cut off from the stimulus of the outside world—they would often develop the habit of walking round and round in circles, placing their feet in exactly the same places each time. So precise is this repetitive behaviour, that in some of these old pits one can see where the animals' endless pacing has worn the ground smooth, until a polished circle has been made in the concrete.

Even the most unlikely species can behave in this stereotyped fashion. We have watched a sea-lion in a zoo in northern England swimming round and round its pool, turning on to its back at a precise spot at one end of its pool and righting itself at another specific spot at the other end. The identical pattern has been observed in a wildlife park also in the north, two zoos in the Midlands and at least two in southern England, clearly indicating that a number of sea-lion pools are unsuited to their occupants.

Another form of stereotyped behaviour which is more often seen in human beings than animals is the ritualized movement of the head or part of the face. One major zoo has a polar bear which swims to a certain point on the outer wall of the pool, pauses and throws its head sharply back and round in an anti-clockwise direction, then immediately swims away. Within a minute it will be back again to repeat the procedure and it does so continuously. In another zoo in southern England we saw a puma pacing a typical old-fashioned cage and executing a similar movement. Pacing up and down in the manner of the Siberian tiger we mentioned, this animal had elaborated the turning procedure by placing its forepaws on the wall and rotating its head each time it reached the end of the cage.

To begin to understand why captivity can distort behaviour in this obscene way it is necessary to return to a point raised earlier: that some behaviour is instinctive and some learned. Most people are familiar with the terms instinctive or innate behaviour and learned behaviour. But how greatly do these factors affect the animal upon which they operate?

Innate behaviour might be regarded as a conditioned pattern that comes into operation as a result of stimulus. If the stimulus is absent the pattern will remain dormant, but once the stimulus or trigger mechanism is presented, impulses from the brain set the behaviour pattern into motion. The lower we descend the evolutionary ladder, the more innate or instinctive is the behaviour of the animal. Alternatively, as we ascend to the monkeys and apes, so this behaviour becomes more submerged by the actions of learned behaviour. Distinction between the two is of great importance, for both have their particular advantages and weaknesses, but it is extremely difficult to draw a definitive line between instinct and intelligence. Ethological research continually unearths new marvels that cause surprise and even shock to the scientific establishment and seem to defy the scientific desire for categorizing and pigeon-holing. Lowly creatures like the octopus have, in recent years, been found to be far more reasoning and intelligent than was expected and in laboratory experiments even the earthworm has been found capable of learning, although admittedly rather slowly! Innate behaviour is passed from parent to offspring, generation after generation; even if a number of generations are denied the opportunity of fulfilling these patterns, the next generation to be provided with the correct stimulus will immediately fulfil the action.

It is impossible to proceed from this point without citing Eugène Marais' unique experiment with South African weaver birds. These attractive little birds plait a tiny little nest on the end of a thin branch. It is made out of grasses and is literally tied into place. The adult birds are seed-eaters, yet the chicks are fed exclusively on worms, prior to leaving the nest. Marais took some weaver bird eggs and hatched them under canaries. He did this for four generations during which time the birds were denied all access to worms, insects and nest-building materials. With the fourth generation he provided all the materials that the birds would find in their natural environment and, as the reader may already have guessed, they plaited a number of nests and when the chicks hatched, the parent birds fed them on worms!

Further up the evolutionary ladder, we start to discover a new type of mentality. Innate behaviour still predominates, yet the animal shows an ability to learn and in many species such learning is a vital part of its behavioural repertoire. Carnivores are of interest in this context. At this level of development innate behaviour is still the governing influence, yet this alone does not seem to suffice; a period of elementary education is also necessary. Hunting shows this quite clearly, particularly in social species like the hunting dog and the lion. Those who are acquainted with the books by Joy Adamson about the lioness *Elsa* will already be familiar with this phenomenon. Although perfectly equipped to kill, the lion still needs to be trained to do so. However, it must be emphasized that this training is of a strictly limited nature. The cubs do not copy the behaviour of the adults. Adult animals endeavour to make prey accessible to the cubs so that they can develop their skill; this skill they eventually perfect on the basis of their own experience, generally acquired after many abortive attempts at a kill.

Ultimately we come upon those species that evidence a greater degree of learned rather than innate behaviour, namely the primates. 'Intelligence' becomes more advanced as one proceeds from the lower primates towards the baboons and the anthropoid apes, where it reaches its peak, second only to the learned behaviour of man.

Earlier we said that both types of behaviour have their advantages and disadvantages and, before proceeding to the respective problems that each can present in captivity, it might prove profitable to look at their individual weaknesses in the wild state. As so many scientists deem it necessary to make a complex subject even more obscure by the use of forbidding terms to describe specific facts, we shall turn again to the works of Eugène Marais and borrow the simple yet descriptive nomenclature preferred by him. Marais described the two types of mentality as phyletic and causal, the former describes the inherited memory and the latter the type of memory that relates to cause and effect.

Those species that are governed by phyletic memory (innate behaviour) are completely equipped both physically and mentally for survival in a specific environment. They instinctively know what is eaten, how to select shelter, even when to migrate and reproduce. Marais' weaver birds did not require nest-building lessons, nor was a course on infant-feeding necessary, even after a forced-behaviour abstinence of four generations.

A combination of this type of psychological and physiological equipment initially appears to be flawless, providing, as it must, an assurance that the species will survive. Unfortunately this is not so, for although both the behavioural and physical equipment can be modified by the pressures of natural selection, the process takes time, something that man has already stolen from nature.

In the face of a sudden environmental change, the animal with a predominately phyletic memory is therefore at a distinct disadvantage—its innate responses can actually create a barrier between the species and survival. Creatures of the forests, having lost their natural habitat, will often find it impossible to adapt to more exposed countryside and vice versa. Even though there might be alternative food sources available, they are psychologically unable to exploit them. A change in life-style of such dramatic proportions would necessarily be gradual; it certainly could not be encompassed by one or two generations. On the other hand, those species that evidence the causal memory (or learn by experience) can to a greater degree free themselves from the restrictions of being tied to a specific environment.

But captivity can reverse these advantages, as Marais discovered. He had a wonderful opportunity to observe the distinctions between these behaviour patterns with two animals, namely an otter and a baboon. Both animals were newly born when they came into his possession: the otter was suckled by a bitch and Marais fed the baboon on a bottle. He kept them for three years, during which time they were fed on food which was not their natural diet. The otter did not see water apart from that in its water dish and the baboon was denied contact with wild fruits and insects, usual ingredients of its natural diet.

At the age of three, the otter was taken back to its natural home and had its first sight of a river. After a few moments of hesitation, it dived into the water and within 30 minutes it had caught and eaten both fish and crustacea. The baboon, when returned to its native habitat, fared less well. It knew nothing of turning over stones in order to find the multitude of insects that hide beneath them and the first wild fruit it picked and ate was of a variety known to be deadly poisonous. Marais, knowing that the baboons of that area were particularly fond of scorpion, persuaded his specimen to taste one that was freshly killed. The baboon obviously found it palatable, but when given the chance to catch a live scorpion, he was promptly stung. Wild baboons do not make such painful mistakes.

If we relate these factors to the zoo we find that those species with positive instinctive drives will only adjust if the conditions provided are in line with their normal way of life.

Arboreal species for instance will never make a total change to life on the ground. This is not just because many are so physically well-adapted to climbing that they are unable to venture down from the trees, but because they instinctively feel insecure if they do so. When one of us was a student keeper at the Jersey Zoo he unwittingly demonstrated this fact when re-branching a cage for an animal called a slow loris—a slow-moving, arboreal primate from the jungles of South-East Asia. Whilst the old branches were taken out of the cage, the loris was placed on the ground, whereupon it immediately began to display the classical symptoms of terror, voiding its bowels and urinating. For a few moments it grovelled helplessly about, then struggled across the floor to the wire front of the cage which it ascended at double its usual speed to the ceiling. Here it remained hanging grimly to the wire until the new branches were fixed into position. Once this was done, the animal soon summoned up enough courage to explore its new surroundings and sniffed and peered about, gradually becoming calmer all the time, and eventually settling down to sleep in its customary curled-up position on a conveniently forked branch.

Animals with rigid drives of this kind cannot learn to overcome them, so they must be provided with a well-branched area that allows them to get well up above the ground.

By comparison, moving up the scale of intelligence, we find that other predominantly arboreal species will show a degree of adaptation that the loris would be quite unable to achieve.

The Jersey Zoo has amongst its unique collection of animals a breeding group of rare ursine or black and white colobus monkeys, a West African species that spends almost its entire life in the topmost branches of the thick forest canopy, hardly ever descending to the ground. The original animals were captured by the zoo's honorary director, Gerald Durrell, in Sierra Leone, and on arrival in Jersey were provided with a spacious, well-branched cage that seemed to suit them admirably. However, the zoo authorities decided that as colobus are so definitely an arboreal species, it would be greatly improved if the monkeys were able to climb higher, so the cage was raised to a height of about 20 feet.

Largely as a result of the natural dimensions of their environment, the Jersey colobus are decidedly more relaxed than many primates seen in zoos and they seem to have greatly overcome their 'floor-fear', often descending to the ground and approaching the public in an almost cheeky manner. If suddenly frightened, they 'take to the trees' of course, but this is a normal, instinctive reaction, not a neurotic captive response. Although these monkeys are kept in a good environment in one of Britain's best zoos, the significant point here is that they have actually surmounted the psychological obstacle of descending to the floor, something the loris could never do.

But, as instinct gives way to learning, the animal is forced to make sacrifices that under normal conditions would be worthwhile but in captivity may be disadvantageous. This is true for many species, but the monkeys and apes are the most pitiful victims and can suffer worse than most animals in captivity. For all of the higher primates the most important periods of the individual's life is that from birth to adolescence, a time of fun and games in which the youngster receives its basic education in social customs and learns the warmth of family bonds. Laboratory experiments with monkeys have clearly proved that babies removed from their mothers in the first few weeks of life, even for only a few days, will grow up to become socially deprived and often make poor mothers themselves. Yet it has often been the practice of zoos and animal-dealers to import newly captured babies to form the nucleus of zoo exhibits. It must follow that a great number of these groups start off with psychologically abnormal animals and this, added to the fact that a lot of zoos keep these animals in impoverished conditions and in unnatural sex ratios, produces exhibits that are little more than lunatic asylums.

Previous theories on the natural behaviour of baboons, for example, had been established through studies made mostly in zoos, but in recent years the excellent field studies made of these animals have shown earlier views to be completely false. In one well-documented captive study, not only were baboons kept in overcrowded conditions, but there was also an unnaturally large number of males and therefore fighting was serious. The researcher concluded from this that baboon society was built on sex and jealousy and that continual struggles for dominance occurred. Nothing could be further from the truth and this document can now really be considered a paper on stress in captive monkeys due to overcrowding and abnormal sex ratios!

Zoos and safari parks seem slow to learn from these mistakes, continuing to duplicate them in spite of the abundance of information available on these species. In the late 'sixties and early 'seventies the safari park operators imported species like baboons and rhesus macaques by the crateful. Regardless of sex or age these creatures were shipped into Britain to stock the so-called monkey jungles.

Despite their deceptively 'natural' appearance, these monkey jungles are remarkably unsophisticated structures, being really nothing more than a fenced-off area of trees and grass with a road through the middle. This is not meant to imply an absolute condemnation of the concept, merely its execution. Sex ratios are often unbalanced and seem to depend far more on the job lot purchased than a careful selection of suitable specimens, and often species are mixed quite indiscriminately. One park we visited had a female hamadryas and two olive baboons mixed with a group about 70 anubis baboons, despite the fact that these species not only inhabit significantly different areas, and that their normal social grouping and behaviour are distinctly different.

The safari park operator might argue that the area provided will do much to dispel aggression. The baboons can get away from each other if they wish, which is more than can be said for these animals when kept in traditional zoos. It's a doubtful argument. British winters are often severe and at such times the baboons require shelter and warmth and they will huddle together in order to get this. Then, the area provided is of far less significance than the quality and quantity of the shelters provided, for the all-important area becomes the far smaller sleeping place. From studies of wild baboons it has been found that at night they congregate in groups varying in number, but entirely dependent on the personal choice of individuals. Thus 100 baboons might split into uneven groups of three or five, six or even ten. In order to cope with the personal likes and dislikes of the individual, it is essential to provide a surplus number of sleeping quarters, so that should A not wish to sleep with B, it can sleep elsewhere without having to resort to a night of exposure in an English field.

None of the monkey jungles visited by the authors seem to conform to these ideas and we have yet to see what might be considered adequate shelter. One park provided four wooden kennels a little over a metre high to accommodate about 70 animals; another, the concrete bunker mentioned in an earlier chapter, made from sewerage pipes buried into the ground. The latter are dark even during daylight hours and in the winter months water seeps into them so they become damp and inhospitable. When the parks close and darkness begins to fall, the baboons move into these sleeping areas; then the aggression increases and fighting starts and often a number of animals are forced to sleep outside. The stress caused is quite unnecessary and would easily be eliminated if more, and better, sleeping boxes were provided.

Conditions in the majority of these 'jungles' are clearly very abnormal. For example, visitors might note the considerable number of baboons with short or stumpy tails and wonder why this is so. To be perfectly honest, we cannot provide a hard and fast answer. Some authorities think that the tail is lost because of frostbite; some put it down to fighting and some think it caused by self-mutilation. Possibly in the case of the safari park it is attributable to all these factors. What we do know is that once a monkey obtains a lesion or some form of injury to its tail, it will invariably lose the piece from the site of the injury to the tip. This is because the blood-flow through the tail is limited, and a cut will cause the veins to seal off so that the end of the tail becomes deprived of blood, dries up and drops off, but although the process is easily understood the cause is often difficult to determine. Nevertheless, we feel that stress, be it due to abnormal social grouping, overcrowding, or lack of sufficient sleeping accommodation, exists quite clearly in these exhibits and is quite needless.

The worst incident we came across was in a collection with concrete bunkers. We asked a keeper on the section how he could safely enter these to clean. As it was impossible to determine whether any baboons remained inside, the risk of being bitten was quite strong. By way of explanation he produced a Thunder-flash (an extremely potent type of firework) and then proceeded to give us a demonstration of the method by lighting the fuse and hurling the explosive into the bunker. As it detonated, half a dozen terrified baboons hurtled out of the exit, shrieking in absolute horror at the sudden and violent intrusion on their afternoon sleep. The method typifies the attitude.

At the other extreme one has the commercial zoo with tiny, bleak cages containing monkeys in two's and three's, or often in solitary confinement. The price paid by the inmates is even more terrible than the previous example. Under the extreme pressures of isolation, behaviour misfires in the most disgusting ways. Grooming, a normal part of the social life of many primates, develops into a fetish. They pluck each other bald and pull flesh until it bleeds. Some, shut off from normal companionships, eat their tails, mutilate their arms, chew their fingers off. Others develop a form of hypersexuality, the sexual response becomes abnormally increased until masturbation and mating seem to govern the animal's existence. These are similar patterns of stereotype behaviour. The cause is basically a decreased involvement with the outside environment.

Abnormal grooming behaviour is commonplace in many

animals in captivity and is particularly dangerous and objectionable because it will cause a hairless, or featherless state that will often lead to open sores and self-mutilation. Of course, grooming is a normal displacement activity in the wild but, in animals that are not occupied with seeking food or stimulated with an interesting environment, it can become a bad and obsessive habit. True, there are other causes for self-grooming, such as parasites or skin diseases, which may act as the stimulus that initiates the sequence, but whatever the cause in any specific case, it is not an easy habit to correct once firmly established. Most often seen in primates and some species of birds, such as the parrot family, excessive grooming has also been seen in various cats, members of the dog family, bears, and a number of other species, though not nearly so commonly. In one wildlife park alone we have seen a pair of scarlet macaws in a tiny cage, absolutely naked save for some feathers on their heads, tail and wings. We saw a cherry-crowned mangabey eating its tail because it was kept in a cage, on its own, that was completely empty. Not one single object of therapeutic value was provided. A large male mandrill in similar conditions had chewed part of its hand off in its frustration. Tragically these are not isolated incidents.

Good zoos encounter these problems less frequently and we were interested to hear from the Director of Twycross Zoo in Warwickshire (a collection specializing in primates) that they had never experienced a case of self-mutilation. Caging here is good, for most of the primates have reasonably large, grassy areas filled with a profusion of branches, and they are kept in biologically appropriate sex ratios. Even so, this cannot be the whole answer. From observations made at Twycross we feel that a large part of its success with these animals is attributable to the constant involvement between keeping staff and their charges.

The natural environment is never completely static. There is always something new. But in captivity the novelties are gone, all is familiar and one of the main preoccupations of life—finding food—has been removed. This can give rise to behavioural problems in most animals, but primates, with their greater degree of intelligence and constant curiosity will quickly become bored and morose. At Twycross the monkeys' day is enriched with stimulants like leafy branches to play with, two or three feeds daily, plus a considerable amount of attention from the attendants, who spend the bulk of their time with their own particular section, rather than moving from one to another. To achieve this degree of involvement the zoo has a fairly high number of staff in relation to its size—a high number, that is, by comparison with many other collections of a similar size. But it is not Twycross that is over-staffed, rather that many others often have too few keepers to cater for the demands of the animals.

Of the great apes—orang-utan, gorilla and chimpanzee—the chimpanzee is the greatest victim of neglect, primarily because it is generally the top-class collections which exhibit the other two species whereas the chimpanzee is kept by a wide range of institutions, from the quality zoo to the circus. Abnormal behaviour in chimpanzees is displayed in various ways. Coprophagy (eating of faeces) is a common problem that seems to have something to do with an increased interest in self, due to boredom. For many birds and some species of mammal, for example the rabbit, it is a normal habit and it is not infrequently seen in the dog family, although usually in juveniles, in which case it is probably associated with growth and digestion. But it is abnormal when found in animals like chimpanzees and orang-utans. It has been suggested that the habit is usually acquired by animals that spend most of their lives in the trees, for when they defecate the faeces drops through the leaves to the ground below and is never actually seen by the animal itself. Therefore, it is reasonable to expect that this unknown material might be of interest to a captive specimen. Certainly the theory seems valid in relation to the orang-utan because these apes spend nearly all the time in the trees and in zoos they are prone to develop the coprophagic habit.

Gorillas, on the other hand, are ground dwellers that would come into contact with their faeces in the wild. In zoos they are rarely known to eat it. The theory begins to fall apart when other species of monkeys and ape are considered, however. We have not observed it in colobus monkeys, despite their being totally arboreal. Possibly this is because they are less intelligent, and perhaps because their faeces are not sticky and pliable like that of the orang-utan and chimpanzee, but harder and of a fibrous texture that makes them less interesting to play with. In addition, colobus are extremely difficult to keep in captivity, so usually only the best collections exhibit them and keep good social groupings in spacious enclosures. Abnormal behaviour in these monkeys is almost unheard of for this reason.

The chimpanzee also throws the arboreal theory on coprophagy. They are neither totally committed to life in the trees or on the ground, yet we have often seen them eating their excreta in zoos. The background is always the same. The chimpanzee starts out by playing with its excrement, moulding it as though it were clay or plastering the walls with it. As some of the faeces adheres to the fingers it is not surprising that eventually the ape licks it off and gradually the less savoury habit develops.

Boredom is a continual threat to the mental health of intelligent creatures like the apes, so their environment must be filled with as much interesting material as possible if they are not to succumb and become mentally atrophied. It has been shown under test conditions that rats raised in impoverished environments show a marked reduction in brain size, and a reduced learning ability. Relate this to the ape and one starts to find similar evidence but on a far greater scale—they become either moronic or psychotic.

In recent years we have been extremely disappointed in many of the seemingly modern and architect-designed ape houses, because if one removed the acres of plate glass and the butterfly roofs, the structure remaining would prove to be unimaginative as far as the inhabitants are concerned. The cage furnishings are primitive and the all-important factor of animal husbandry backing up the exhibit seems of a poor standard.

A famous, long established zoo did develop one of these enclosures. On three separate visits we watched a gorilla housed in it, walking mindlessly round in circles, bashing the same square of plate glass with its fist at each revolution, simply bored to the point of insanity. Another zoo, lacking the finance and pretentiousness of the first keeps its chimpanzees in a square, barred cell, with only an old car tyre hanging from a rusty chain as amusement. At the time of our last visit to the collection, in 1975, these pitiful wretches sat huddled together in abject misery on a floor coated with faeces and rotten fruit. The male spent the whole time obsessively, but unenthusiastically, masturbating.

At Twycross, one can see the same species exhibited in fairly conventional cages, but totally occupied, playing with companions, stimulated by the attentions paid by dedicated keepers and seemingly enjoying life to the full. This is true of all the best zoos. Basle in Switzerland, the first European zoo to breed the lowland gorilla, has an enclosure for this species that includes various mounds, shelves, tunnels and climbing frames, and the gorillas are kept together in a family group. Even the cleaning of the caging is part of the daily ritual and is contrived to involve the apes with the keepers. Cages are cleaned as is deemed necessary (this can be several times a day) partly to maintain a high standard of hygiene and partly to avoid any chance of the development of coprophagic habits. Most cages can be cleaned whilst the apes are in residence and it is interesting to see how much the gorillas love the water that is sprayed from a high pressure hose on to the floor. They even encourage the keeper to squirt them with it. Water is not in any way detrimental to apes, in fact the Basel specimens have beautifully soft skin and a thick pelage that is quite exceptional in captive apes.

To return to Twycross in England, we find that the gorillas here are also provided with water in the form of a sunken pool in the interior quarters, and they have other amusements, including a colour television. Hardly a natural diversion, yet it is a genuine attempt to give added stimulus to the lives of the apes and is unlikely to prove harmful in any way. As a point of interest, this experiment is not entirely successful. The Twycross gorillas are fairly impartial to the offerings of television and only seem to really enjoy animated cartoons and wrestling! On a slightly more serious note, all the apes at Twycross are fed three times daily and are given wood wool which helps satisfy their normal nest-building activities and provides them with a clean and comfortable bed. Twigs and branches are supplied for the animals to eat and play with, and the keeping staff are almost always in attendance.

In 1974, when the new orang-utan enclosure was completed, the directors spent a considerable amount of money on having three trees, approximately 40 feet high, moved to the grassed outside enclosure because they felt that the apes would not only enjoy the climbing facilities, but also 'such a pleasant view of the countryside'. Of all the ape-houses in Britain this is, in our opinion, perhaps the finest of all.

The meticulous attention to detail with regard to the primates is a sure indication of the advantages of zoos specializing with specific species, for the standards of the primate collection in this zoo are way above those of the other exhibits to be seen here. Not that the remainder are in any way reprehensible, in fact buildings like the reptile and giraffe houses are very fine indeed, and far better than most others in this country. But they seem to be overshadowed by the incredible variety of monkeys and apes that live in so contented a way in such excellent conditions.

In this chapter we have touched upon only a few examples of ways in which behaviour can misfire in captivity, but hopefully enough for the reader to appreciate that it is the duty of every wild animal collection to do its utmost to neutralize these adverse effects. In the light of present-day knowledge, this is not as impossible as it may initially appear, and all of the problems cited are capable of being solved, as they are basically ones of animal management, some more complex than others, but the majority just plain common-sense.

It has been firmly established that every creature is a prisoner in space and time, restricted by environment, conspecifics (members of its own species) and relationships with other species. Each has its home range which, when protected, is called its territory and it is not free to wander outside it as this would usually mean encroaching on the preserve of another.

The home range or territory is not of uniform value to the animal because certain areas are used for different functions. Feeding is a very important one, occupying most of the animal's time. Water will be found elsewhere and the animal will have certain places for sleeping, hiding or sunbathing.

Animals do not occupy an area as a piece of real estate. Rather its size is dictated by food supply and the area becomes larger when food is scarce, but contracts when food is plentiful. The area is bounded by natural or imaginary boundaries and is occupied by other species which add a profusion of stimuli to the animal's life. It is thus dependent not only on the area but on the quality and variety of the environment and on the structure of the family and social group.

These factors can to a certain extent be transported into a zoo. Once the animal has been furnished with its personal possessions, in the form of suitable cover, a tree to mark, a sunning platform, and sandpit, a bath, or whatever else its biological needs dictate, it will then assume its own home range or territory. Under good conditions many wild animals will become extremely settled and contented with their artificial environment, and it is well-known that many escaped specimens will return quite freely to the security of their enclosure. But it is also known that animals and indeed man will return to the environment he knows, however bad—within reason, that is—rather than brave the unknown. So this action does not necessarily mean that the conditions are good.

What we are looking for in the zoological collection is simply an acceptance of the undeniable and obvious fact that all the inhabitants have sense organs, behaviour patterns and feelings. It is virtually impossible to know how an animal feels subjectively when it is sad, angry or happy, and ethologists frown on attempts to interpret animal behaviour in this way because anthropomorphism (looking at the animals' situation from the human viewpoint) invariably distorts the true facts. Nevertheless, a sight more compassion and understanding would not be amiss in the majority of our British wildlife collections: an element of tenderness expressed in the erection of cages and enclosures that contain stimulus and variety; animals kept in families or social groups that correspond to those found under natural conditions so they can establish normal relationships and escape the hell of either solitary confinement, or the unrelieved tensions of over-population.

We would also like to see the employment of staff who are ably suited to the arduous task of caring for these often extremely demanding creatures, not just cheap local labour, but people with a proper training in all aspects of animal husbandry. If this were so, many stupid errors would cease to be made. Errors like placing the feeding dish of an arboreal animal upon the ground, or a nervous animal's food too near the public, so making it psychologically difficult for the creature to eat in comfort. Or dumping food and water for a herd or group in one spot so that the dominant members take all. Or cleaning an environment so thoroughly that the animal's scent or 'identity card' is removed, forcing the inhabitant to frantically re-mark its territory day after day until the ritual becomes a stereotype pattern (we will deal with this factor in more detail in thc following chapter).

It would also be refreshing to see more of the British zoo-owners motivated by an interest in zoology instead of profit. For behind every good zoo one will find a director who is genuinely interested and is not ashamed to use the word love when speaking of his charges.

Finally, a greater degree of honesty and responsibility is needed. Some species find the conditions of captivity almost impossible to adapt to, so it is time that wildlife institutions either reconsidered the present methods of maintenance, or made a positive decision not to keep those animals until certain that it can be achieved without causing suffering. Elephants are, in our view, a borderline case. They are difficult to keep owing to their size, sheer strength and intelligence and even the costliest housing provided seems to be inadequate. All the U.K. elephant exhibits are too small and the long-established practice of chaining these creatures at night often leads to the stereotyped actions of weaving from side to side, nodding and shuffling back and forth in a ritualized way. Safari parks present a better image by showing their elephants in grassland, but then the general public are probably unaware that these animals, too, spend their nights chained in houses that are sometimes less suitable than the standard zoo type. We therefore feel that the keeping of elephants in captivity needs a fresh approach. It would mean a number of spacious areas for daytime use, large sleeping quarters, possibly with automatic doors for staff safety, indoor heated pools, breeding pens. The cost of such a project would be enormous.

And the same criticisms must apply to the keeping of dolphins. Dolphinariums are currently the top zoological attractions, a regrettable popularity that has resulted in the worst commercial exploitation in the history of British zoos.

Dolphins require a pool that provides a physically and acoustically comfortable environment: rectangular or circular tanks with vertical walls will reflect sounds emitted by the animals, causing distress and disorientation. Of course, the dolphinarium operator will ridicule such criticisms and one has stated that noise pollution does not exist because the animals do not talk and use their sonar system all the time.

But in October 1972, Mr Farooq Hussain, a biophysics researcher from the University of London, stated in a report made to the Universities Federation of Animal Welfare that 'All the existing pools in which *Cetacea* are kept in captivity in the United Kingdom are inappropriate in both design and material construction'. The following February he added in an article in the *Sunday Mirror*: 'I have researched the figures for British Dolphinaria, and I believe that about 20 per cent die every year.' In the same article two other experts on dolphins endorsed Hussain's comments. Dr Van Heel, Founder of the European Association of Marine Mammals, severely criticised the conditions in British exhibits, apart from Whipsnade's which he thought very good, and added, 'I have no death figures, but 20 per cent might be conservative'. And Dr John Lilly, who was the pioneer of human communication with dolphins, made the gravest condemnation of all. He suggested that because dolphins are so intelligent they get bored with doing the same circus tricks day after day and will often be driven to committing suicide. Lilly maintains that six of his dolphins did this so he turned the rest loose in the sea and after 12 years of working with these animals he abandoned his researches. His comment was 'I felt that I had no right to hold dolphins in concentration camps for my scientific convenience'. This is the type of moral decision that distinguishes the animal lover from the commercial exploiter.

The fears of those of us who are concerned with the legitimacy of the zoo as a cultural institution were well expressed by Dr Desmond Morris, author of *The Naked Ape*, who for 8 years was Curator of Mammals at London Zoo. In an article that appeared in *Life* Magazine in 1968 he wrote:

> A city zoo normally covers a small area and houses a wide selection of species—something of everything. This is essential, it is argued, in order to attract enough visitors to pay for the upkeep. Cram the animals in and give the public its money's worth. This, I submit, is a major flaw in zoo thinking.
>
> I myself, as a zoologist, have been guilty of subscribing to this approach in the past, but for different reasons. For me and many other zoologists it is fascinating to see large numbers of species side by side for comparative purposes. But after studying the results of this policy in any normal-sized zoo, one is forced to think again. If this policy leads, as it seems it must, to the exhibition of frustrated, bored or neurotic animals, then the day will come when zoos will nauseate our society.

Chapter 7

# Zoo husbandry

Some years ago there was a zoo keeper who had amongst his many charges a two-toed sloth. As most people know, the sloth lives in trees and spends most of its time climbing and hanging upside down. It is not a creature renowned for its activity. Indeed, so slow-moving is the animal, that its matted hair usually has algae growing in it and, in the wild, its fur is nearly always infested with certain small moths, which are found only on sloths. In short, it is an animal that exerts itself as little as possible.

This particular sloth lived in a small cage provided with a couple of stout branches upon one of which it remained suspended fast asleep for the duration of the day (sloths are nocturnal), so the keeper saw little of the animal in action. Each day he entered the cage, swept and washed the floor, placed fresh fruit and vegetables in the sloth's bowl and changed the water, then departed to deal with the other animals in his care. Then one morning as he was sweeping the cage he accidently struck the sloth with the broom handle, the animal crashed to the floor in a heap and hundreds of maggots issued forth from its body on to the concrete—the sloth had been dead for days.

We cannot pretend to have witnessed this incident, but we have little reason to doubt its authenticity, as the story was related to us by a zoo director of impeccable reputation to illustrate the problems of obtaining good staff. Nor do we doubt the fact that the dead sloth remained hanging on the branch. We have ourselves tended a sick two-toed sloth which, despite its weakness, managed to retain a tenacious grip on its branch with its hook-shaped claws. When regrettably it died, it did fall to the ground, but not immediately, so it does seem feasible that a sloth could remain in the hanging position for some time after death.

We deem it relevant to recount the tale because it does, as the director suggested, highlight the kind of thing that happens when zoo staff are of a low calibre. Just think: this keeper cleaned, fed and watered this cage daily, obviously oblivious to the sloth contained in it, and not once did it strike him as odd that the food and water remained untouched, or that there were no faeces on the cage floor. Incredible it may be, but unfortunately it is not untypical of the kind of failings found in many zoo personnel.

We have encountered similar cases ourselves. A zebra in a British safari park was evidently off-colour, its eyes were sunken and its coat coarse and dull, yet the previous day the animal had seemed to be in good health. When the keeper, or ranger as they prefer to be known as in these places, was asked if he had any ideas as to the cause of the zebra's sudden decline, he pondered for several moments and then suggested that it might have been because she hadn't eaten for a couple of days. It had not occurred to him to report this to his superiors and in fact on the daily report each section was supposed to submit to the main office, he had written 'O.K.' alongside this particular animal's name. The zebra died later that day of a blocked bowel. Had action been taken two days earlier, its life could probably have been saved.

Possibly the most horrible and idiotic incident we heard of was one concerning three spectacled caimen kept in a small zoo. The animals were housed in a vivarium within the reptile house that was pleasantly furnished with rock plants and a heated pool, and the latter had to be cleaned every week. This cleaning operation was simple enough, the caimen were removed from the pool, which was then emptied, scrubbed clean and refilled with luke-warm water. Luke-warm water was essential as the heating system in the pool took a while to heat cold water to the required temperature, and a sudden chilly immersion could have proved detrimental to these small reptiles.

On the day in question the reptile keeper followed the usual procedure, the caimen were removed from the water and placed among the rocks at the back of the vivarium while he cleaned their pool, then he went to fetch some buckets of water to refill it. It took about four buckets to get the pool to a reasonable depth and this was usually obtained from a nearby food preparation room where hot and cold water were mixed until a temperature of 65°–70°F was reached. But this bright young lad could not be bothered with carefully preparing the water, he just filled two buckets with cold and two with scalding hot water. Unfortunately he took the hot water to the reptile house first. In the brief time it took to return with the cold water all three caimen plunged into the pool and on his return he was greeted with the sight of the caimen, boiled and very dead, floating upon the surface of the steaming pool.

Ignorance, stupidity, neglect, whatever one cares to call it—the question is, how do people like this ever come to be in charge of animals? Unfortunately, the pathetically low standards of animal husbandry that pervade so many of our zoological collections are hardly to be wondered at, for only a fraction of their staff have received any sort of basic training. For many zoos it is common practice to hire young, inexperienced people and put them directly to work in the collection, their only qualification for the job being a so-called love of animals. After a few months those recruits find that they are merely entrenching themselves into a job with limited prospects, other than damned hard physical work, and their dreams of job satisfaction begin to fade rapidly; many leave during the first year. The constant drifting of 18 to 20 year-olds into and then out of our zoos must be considered a serious indictment on the professional standards of these collections.

The best (or worst, depending upon how one looks at it) examples of this we have come across have been in safari parks and some of the smaller new zoos. Some safari parks seem to employ virtually anyone willing to accept the modest wages offered, the most singular distinction made being between applicants who can drive and those who cannot. If unable to drive, a ranger will most likely be relegated to a gate job, be provided with a shotgun, walkie-talkie radio and a game ranger's uniform, and from this point onwards his daily routine will consist of opening gates for visitors' cars and brewing tea for the rangers in the section. If the applicant can drive he might be given a tractor or landrover, in which he will spend the bulk of his day, at a place appointed by his head ranger, on the look-out for visitors opening their car windows, or breakdowns. At peak periods he will in fact spend most of his day towing cars that have broken down to areas of safety. Involvement with the animals is limited to mucking out sleeping quarters (not always a daily duty in safari parks), pushing animals to areas where they can be better seen by the public and rounding them up in the evening. With such job satisfaction to be gleaned, it is small wonder that the level of intelligence amongst the bulk of safari park employees is somewhat low!

When the writer applied for a senior post in a safari park, the usual preliminary interviews was reduced to a five-minute chat with the general manager, no references were taken up, he was engaged on the spot and asked to start the following week. So clearly safari parks are not greatly concerned about the quality of the staff they employ. One wildlife park we know was only interested in employing men capable of hard labour, and time spent making careful observations of the animals, a vital duty for every keeper, was considered wasted time by the park owner.

The time is long overdue for the keeper's job to be clearly understood by all zoo owners, for the conventional image of a man with a brush and shovel is as antiquated as the concept of the menagerie. A zoo keeper needs a wide range of knowledge and a comparable range of talents if he is to fulfil his duties adequately; the talents might be innate but the knowledge has to be acquired through proper training.

It is not enough for the animal keeper to know what food, and how much to give an animal. He should understand that feeding involves nutritional value, quantity, quality and texture. If a particular aspect of behaviour is seen in a species, he should understand why. He should be familiar with its country of origin, its habitat, its status in the wild. A knowledge of veterinary procedure is also needed—the control of parasites, treatment of sick or injured specimens (*always* under the direction of a veterinary surgeon), restraint techniques and aftercare. He should also understand the objectives of research, education and conservation, for without objectives he is unlikely to achieve the satisfaction the job can provide. Above all, he should be able to establish a relationship with the animals in his care, not a sentimentalist's overbearing relationship, but one based on trust, man for animal, animal for man. If he accomplishes all this he can consider himself a keeper of some value.

For many the realization of this standard is quite out of the question; the young enthusiast who genuinely wants to work with wild animals might reasonably expect to find the hours long, weekends included, but he will also almost inevitably face the prospect of extremely poor pay and, in the light of many present-day attitudes, little hope of a suitable apprenticeship.

Self-education is a remote possibility. Most books on specialist subjects are expensive and therefore the majority are beyond the means of the lowly-paid zoo keeper and, in our experience, few zoos have a library for the staff to refer to. We have encountered several zoo owners and managers whose personal reference books numbered fewer than 20, the bulk of these consisting of children's encyclopaedias. It may interest the reader to know that one of the only animal encyclopaedias of any use to the zoo biologist is published in thirteen volumes. In 1977 it was retailing at about £260 a set and even this monumental work requires a great deal of supplementary literature to make it of practical use.

Crass as the suggestion might seem, it is our opinion that a lot of zoo owners, lacking the desire to know their subjects themselves, find it preferable to employ staff who are equally ill-informed, perhaps because it would be acutely embarrassing to have staff more able and qualified than the director! A colleague of ours who applied for a senior zoo post was actually turned down because of his experience. He keeps the letter of rejection framed upon his wall as a memento of zoo stupidity.

If this situation prevails and the zoo keeper continues to be looked upon as a manual labourer working with animals, little progress is likely to be made in the field of zoo biology. Staff must be trained to a high standard and, when trained, paid a salary appropriate to their skills. This last point is vital.

In January 1974 we met a keeper, a single girl, with 15 years' zoo experience. Her unquestionable knowledge of the species in her care and devotion to their well-being made her an extremely valuable member of staff, yet she was being paid £17 a week, plus the use of a caravan. We hear much about the plight of the factory-worker, the miner and the docker, but never of the poorly-paid zoo keeper. There are three reasons for this silence. In the first place, the zoo worker specifically chooses to work with wild animals and is therefore expected to obtain reward from 'job satisfaction'. Apparently they are not supposed to need money for the things the rest of society deem necessary. Secondly, they are considered unskilled, which most of them are, simply because few zoos give them a proper training. And thirdly, and perhaps most significantly, how can they protest? They can hardly go on strike because in a zoo dispute between staff and management a third party becomes the innocent victims—the animals. No self-respecting zoo man would allow his charges to suffer, and certainly not because of his own actions. If he did he would not be of the type suitable for zoo employment, so the totally mercenary zoo owner has his employees firmly by the throat. But for how long will zoo staff dedicate themselves to a lifetime of hard and often unpleasant work in the knowledge that it is their interest in natural history and their devotion to the animals that is being exploited? Two years seems to be the average, and the majority of zoo owners effectively encourage this turnover with their attitude of 'if they dislike the conditions they can leave—there is always some naïve wildlife enthusiast anxious to step into the vacancy'. This is why there is such an appallingly high turnover of young employees in the British collections.

No respectable zoo director would deny that in the quality of the staff lies the secret of successful zoo operation. The responsibility for the welfare of individual specimens rests with the keeper and the curator or director must rely almost entirely on their reports, so the keeper is the eyes and ears of park management. Indeed, if the collection is well organized and medical, history and behaviour records are kept—as they should always be—the bulk of information contained in the records must come from the keeping staff. For this to be of any value they must know what to look for in the first instance.

A few of the better collections do actively pursue training schemes for their staff and have libraries to which the keepers have access, but the Zoological Society of London, in conjunction with the Paddington Technical College, provides the only fully comprehensive course on Zoo Animal Management for its own staff. The Zoo Animal Management Ordinary Certificate course covers the basic areas that keepers should be conversant with and includes instruction on kinds of zoological collections, conservation, zoo management structure, safety, classification of animals, cages, enclosures and environment, instinct, drive and learning, health and hygiene, biological classification of foods, nutrition, quarantine, injuries, poisoning, parasitic and biological infections. Such a curriculum should be available on a national basis and it ought to be compulsory for staff to reach a set standard before being placed in charge of their own sections. (It would also be interesting to see how many zoo owners could cope with the course.) But how this could be effected without the complete co-operation of every collection in the country is a puzzle.

To appreciate what working with wild animals is about there are several aspects of zoo operation that should be explained. As each is a complex subject on its own, we have resorted to brief summaries that are intended to give a general picture of the zoo's operational problems and its objectives.

From the onset, every zoo worker should clearly understand his obligation to the cause of preservation and the problems involved in captive breeding. It must be accepted that the rescue of any species by placing it in the protective care of a zoo falls short of the ideal. Successful conservation of any species must take into account protection of its habitat and other life-forms within that habitat. But those who often use this fact to dismiss the idea of captive breeding as unrealistic and fanciful might care to consider that for many species the alternative to these 'zoo banks' is no preservation at all.

At the time of writing it is estimated that approximately 10 per cent of all species of living mammals are seriously threatened with extinction. This figure seems less trivial when it is converted into approximately 200 distinct species. In all, there are over 1,000 living vertebrates under sentence of death. Already we have lost much. Since the early part of the seventeenth century some 95 species of birds and 40 species of mammals have become extinct. The causes were various, the perpetrator invariably man.

Obviously the ideal way of protecting wildlife is to leave it in its natural environment and ensure that it is not molested or interfered with by outside influences. This is usually easier said than done. The commonest line of defence is to establish a reserve; in this way the flora and fauna are given a tract of land that provides sanctuary against the pressures and ever-growing demands of civilized man, yet leaves the pressures of natural selection intact. Ideal as this situation is, there are too few reserves in the world and, of those that do exist, many are too small, suffer heavily from poaching and lack adequate funds. Furthermore the establishment of a reserve can take years from the date of the first tentative suggestions to the final granting of protection by government. For some species these time delays will spell doom.

So although the wildlife reserve is a better insurance policy than most other alternatives, it is at this point that the possibility of preservation in captivity must be considered, and prefaced with that question. Preserving for what? It is a vitally important question. Whenever one sees a zoo that proudly announces the birth of an orang-utan, it is inevitable that the paradox of the situation strikes home. One looks at this wonderful little creature, a perfect miniature of its round-bellied parents, wondering if this breeding success could ever justify the slaughter of so many in the past, a slaughter that was prompted by the zoo's desire to own them.

To preserve a wild animal it is usually assumed that the eventual aim is to get it to reproduce. This is, however, only part of the task. Preservation in captivity is in no way concerned with domesticating a species. The zoo's function is to act as an asylum for threatened species with the ultimate goal of returning selected specimens to their native home if circumstances arise that are favourable to such a scheme. Such an idea might easily be dismissed as Utopian by many sceptics, but what, one wonders, is their alternative? Anything short of this is quite unacceptable, and although we have far to go before we achieve so complex a scheme of animal husbandry, it must, nevertheless be our eventual goal. Even specimens that are never to be returned to the wild must be maintained as constant with their wild counterpart; this is essential because they constitute part of the captive gene pool, and any abnormality will jeopardize all future efforts to secure the survival of the animals. Thus, the maintenance of the 'type' is of the maximum importance, for only the animal which is representative of the true type species is of any value to research, education or preservation efforts.

Obviously animals in captivity are living in an artificial environment and no matter how good the zoological park is, there are distinct possibilities that the effects of such a situation could drastically alter many natural characteristics of captive specimens. These fundamental changes cause considerable concern to the zoo biologist, as rescue of a species by captive breeding is supposed to lead to the eventual re-introduction of at least some individuals to their places of origin; something that could not be considered if either physical or psychological abnormalities were present in the zoo-bred stock. To add to the problem, if a species has been drastically reduced in numbers and there are only a few individual animals available, the biologist has a limited gene pool with which to work, and if care is not taken a race of degenerate animals will be produced. Under normal conditions there is a constant reshuffling of genes as a result of unrelated animals mating and producing offspring that inherit the stronger attributes of their parents. If the number of unrelated animals is brought down to a level where the gene pool is too small, the variations that enable a species to survive are much reduced, and weak characteristics, such as poor eyesight or hearing, or tendency to disease, are no longer overridden by the stronger features, and inevitably the species will die out.

To avoid unnecessary inter-breeding in captivity, stud books have been compiled on some of the rarer animals, such as the Przewalski horse, Siberian tiger and Père David's deer, making it possible to check on the parentage of a specimen before mating. The advantage of keeping such records and actively co-operating in exchange for the purpose of introducing fresh blood must be obvious, but as many British zoos do not keep comprehensive records of their animals, brothers are mated with sisters, fathers with daughters and so on. The results of this practice can be clearly seen, for example, in the many degenerate lions found in zoological parks throughout the country. Some have drooping mouths, short legs, piggy eyes and many are subject to a peculiar complaint commonly known as 'star-gazing'. The complaint is aptly described by this term, as the affected animal stares skywards with the eyelids seemingly pulled back; sometimes the spine is arched, and the poor creature frequently trips over objects in its path. Various theories as to the cause have been advanced, yet the evidence weighs heavily in favour of inter-breeding and the consequent drawing on a limited gene pool.

However, it is also clear that by placing an animal within the protection of captivity we tend to remove the pressures that are essential for the continuation of the action of natural selection. In other words we risk the total loss of the typical animal. Degeneration of the species is by far the greatest risk to all programmes of captive breeding. Variations of the standard type are generally those that fall before the scythe of environmental pressures in the greatest numbers. Basically they are not so well fitted for the conditions as the normal representatives of the species. Ornithologists have discovered that after severe hurricanes many dead seabirds may be found for miles along the coastline. Of these, by far the greatest number of casualties occur in the birds that are either above or below what is considered the normal size for the species. Examples of this kind are to be found in abundance in the reports of field naturalists throughout the world.

Obviously, if a species is to be rescued by breeding in captivity, its ultimate future must be considered at an early stage of the operation. Is it to remain preserved for all time under artificial conditions, or do we envisage a programme of temporary sanctuary with the eventual aim of re-introduction? The latter is quite definitely preferable, even if it is a more taxing and arduous policy.

It would be impossible to put a time limit on such rescue operations; they might last for five, ten, maybe a hundred generations. Yet the preservationist's purpose must remain the same, namely to preserve the species in an unaltered form from that which is normally found in the wild state. Failure to do so would be disastrous and every re-introduction scheme would be doomed from the start because the animal would lack the equipment needed to gain even a modest foothold in its ancestral home.

The point can be well-illustrated by a classic case of extinction. Menagerie disease is a condition that is of considerable interest to the zoo biologist, as much for its historical associations as its scientific prognosis. It is a condition associated with bears that was much in evidence in captive animals during the last century. It was caused in a similar way to cage paralysis. Bears were usually confined to damp pens and pits with limited opportunity for either physical or mental exercise. Some merely sat in their prisons to rot; others gradually became stereotyped and executed clockwork-type movements until their eventual demise. This restriction of natural exercise will often produce ossification of several of the vertebrae, accompanied by severe inflammation; the spinal column becomes fused and sometimes deformed. Vile as the affliction is, it has a long and fascinating history that contains an attendant cautionary tale, all of which was discovered by the science of palaeontology.

The disease is first found in the bones of the long extinct cave bear and, thanks to the work of the palaeontologists, we have a remarkably clear picture of the life of this animal.

The cave bear was undoubtedly the largest bear that ever lived, averaging between 10 to 12 feet in length and standing at least five feet at the shoulder. It was also incredibly adaptable. A robust creature that was primarily vegetarian, it easily changed to more carnivorous foods if needs dictated, and it was equally able to withstand extreme changes in climate. Unlike other mammals of its age, the cave bear did not migrate according to the changing periods through which it lived, but tried to adapt to the new conditions. For centuries it managed this most successfully. So successful was it that it survived the periods of glaciation without significant loss, yet the species died out completely during the Last Glaciation in Europe. What happened to change its secure position?

The answer is to be found in the success of this huge beast. It was in fact too successful for its own good. Before the Last Glaciation, the cave bear had everything going its way: the summers were long, food of all kinds was readily available and, most important of all, it had few enemies. Sadly, it is this period of abundance that heralded the end for the colossus, for no longer did the bear have to struggle to exist. Life was just too easy, so natural selection ceased to operate. In the absence of natural selection, mutations were allowed to live and so the bears' offspring became short-legged, dwarfed or feeble. Once such a process starts, the species is a condemned one that is no longer able to contend with changes or disturbances in its environment; such was the fate of the cave bear.

With the onset of the Last Glaciation the bear was forced to spend longer and longer sheltering in the confines of caves or fissures in the ground, until eventually it was obliged to spend approximately 8 months of every year under these abominable conditions. It suffered terribly because of it. The bones of these great creatures reveal that they fell prey to inflammation and atrophy of the bones, tuberculosis of the joints, gout, rickets, arthritis and ossification of the vertebrae—all the symptoms of

Menagerie Disease. Forced confinement also allowed the teeth to grow to such extremes that in some instances the mouth became completely blocked and the animals starved to death. Added to these appalling afflictions, the cave bear now had a formidable enemy and competitor, the Neanderthal hunter—a creature who also needed the shelter of the cave for his own survival. Without doubt this 'new' species wrought havoc amongst the now degenerate, sick, crippled cave bears, which would have been particularly vulnerable during hibernation when the Neanderthals would probably have blinded them with flaming torches and stoned them to death.

There are many sites which provide evidence of such slaughter and one of the most famous, the 'Dragon's Cave' in Austria, was the source of a charming mistake. In 1673 a German physician, a keen amateur palaeontologist, discovered a hoard of bones that marked the site of the brutal slaughter of numerous cave bears. In his subsequent reconstruction of these bones, the poor man became somewhat confused and eventually pronounced his find as the fossil remains of dragons! Hence the name Dragon's Cave.

Discoveries of a more serious nature have been made since that time of course, and apart from evidence of the prehistoric bear hunts, there is amazing proof of the part the great cave bear played in the religious life of Neanderthal man. Finds have been made throughout Europe that clearly show that the primitives established collections of bear skulls and in all probability these were spiked on poles and danced around in solemn rituals. Skulls and bones were carefully arranged in crevices and stone chests, and often bones were pushed into the eye sockets. Decapitation of the bears was a common practice that eventually reached fanatical proportions. The wealth of evidence accumulated has led to the supposition that the Neanderthal exterminated the cave bear; but this is in fact rather unlikely. The cave bear was, as can be seen from the evidence of disease and degeneration, an already doomed species by the time of this persecution. At most, the Neanderthal made a contribution to the bear's extinction; there is little to suggest that he caused it.

The cave bear has been aptly described as the animal that domesticated itself, like man. As such it should serve as an object lesson to us all, for the price it paid for its liberation from the pressures of natural selection was death. Nor was its departure from the world without suffering: the last of the cave bears were ravaged by epidemics, diseased females aborted or gave birth prematurely and the few cubs that survived were rickety and arthritic.

For the zoo biologist this horrible tale contains two lessons that are of significance. Symptoms related to those found in the fossil remains of the cave bear have been discovered in captive bears during this century. Unpleasant as the suggestion may be, it is not without grounds to suggest that the Menagerie Disease of the bears is still a factor in many of our zoos; considering the material that is destroyed or thrown away annually (we are referring to the number of deceased specimens that are disposed of without an autopsy), the suggestion might prove difficult to substantiate. There can be no doubt that they are caused by deprivation and neglect. Present-day knowledge of the biological requirements of species should ensure that afflictions of this nature are never again found in any captive specimen.

The second lesson is the disastrous effects that removal from natural selective pressures will invariably have on a species. Every zoological collection would do well to keep this fact firmly in mind and make efforts to avoid methods leading towards domestication. Any creature that escapes the constant rigours of natural selection must ultimately be extremely vulnerable to these same conditions; it will pay in terms of deformity, disease and final oblivion. All zoo employees should be acquainted with these facts as they create the essential framework within which they must work. From this point on, the necessary considerations are of a more basic nature.

Good animal management will only result from a comprehensive analysis of each species and its wild environment to determine into which category the recognizable factors should be placed. A basis for analysing a species in relation to its wild environment might be summarized as follows. Where does the species come from? Considerations should not be on latitude alone, animals that live on the same latitude frequently inhabit different altitudes and are, therefore, exposed to entirely different climatic conditions.

What mode of locomotion is used by the species? Some animals are arboreal, some terrestrial, others make use of both ground and trees. Many species are quite obviously aquatic, seals and hippopotamuses for example, yet water is equally important for species like tigers, tapirs, elephants and even cheetahs.

What does the species eat in the wild and how much of its time is spent eating? The food provided must be of the right bulkiness so that the musculature of the gut works normally, yet it must contain the correct nutrients to maintain the animal in a healthy condition. Carnivores, having eaten their fill, usually spend most of their time lazing or sleeping. Herbivorous animals, however, normally spend the majority of the day eating. Ruminants can be fed a substitute diet quite simply. As we have already mentioned, they eat quickly then retire to a place of security, regurgitate the partially eaten food and chew it at their leisure. The non-ruminants pose a more difficult problem as they need to feed continually throughout the day.

What is the usual social structure? Some species like baboons are gregarious and have a complex social structure; others, like the tiger, polar bear and giant panda are usually solitary creatures except when the female comes into season. Many species of deer form single male family groups, except during the breeding season. And similar multi-male groups occur in pack-living animals like the banded mongoose and the blackbuck. Bachelor herds on occasions develop elaborate ritual behaviour displays and a complex hierarchical structure evolves. To date it is hard to determine the true significance of these hierarchical disputes and without more substantial evidence it is tempting to suppose that once the dominant order is established, preference of females is a mere formality once the breeding season arrives.

What are the criteria that lead to breeding? Baboons will breed under appalling conditions, as incidentally will *Homo sapiens*, who has reproduced within such squalid environments as the Warsaw ghetto, refugee and concentration camps, amidst incredibly oppressive and insanitary surroundings. For other species the requirements are more critical.

Certain species come into season or rut at particular times of the year; this circennical rhythm of daylight length can easily be disturbed by moving an animal thousands of miles to a different hemisphere.

Probably many species require the stimulus of introduction to the opposite sex, rather than constantly living together. They mate more readily after a prolonged period of isolation from their partners. Such appears to be the case with the cheetah and doubtless this rule will be found to be applicable to other animals as time goes by.

With some of the solitary species the presence of the male after parturition can represent a very real danger to the young. Under natural conditions the male would have departed from the scene long before the birth of his offspring. If, under the conditions of captivity this fact is ignored, the male can be extremely aggressive to his mate and her babies, and if he does not directly attack and kill them, the disturbance his aggression can cause will inevitably result in their deaths. The polar bear would provide a typical example of this type of behaviour.

The requirements of the mother during pregnancy must be considered most carefully, because if conditions are not ideal, although the female has been successfully mated, she may abort or, in some species, actually reabsorb the foetus. The period following parturition is equally critical and great care must be taken to see that all the mother's needs are catered for, otherwise she may neglect or even kill her own young. Many factors can cause such disasters, the most obvious being disturbance or fright, but sometimes the true cause can be quite obscure.

Whilst on the subject of breeding, it should be borne in mind that there are already many animals which perpetuate themselves in captivity quite satisfactorily, obviating the need to draw on wild stocks, and many more that could be self-supporting in the 1980s. For example, the following animals breed well in world zoos every year:

| | | | |
|---|---|---|---|
| Barbary sheep | Mouflon | Blackbuck | Yak |
| Nilgai | Eland | Roe deer | Reindeer |
| White-tailed deer | Wapiti | Red deer | Japanese Sika |
| Axis deer | Fallow deer | Guanaco | Llama |
| Wild boar | Hippopotamus | Grant's zebra | Lion |
| Tiger | Leopard | Puma | Brown bear |
| Wolf | Dingo | Sacred baboon | Rhesus monkey |
| Java macaque | Red-necked wallaby | | |

But, as was stated earlier, breeding is only part of the task. Every zoological organization must try to go much further by establishing a total understanding of the physiological and psychological nature of the species in the wild, and then endeavour to maintain these factors in the animals kept within the collection. Preservation in captivity must presume that the resulting stocks will be normal animals as little affected by their isolation from the pressures to be found in the wild as is possible.

Modern methods of feeding wild animals in captivity are formulated on the basic nutritional requirements of certain collective animal groups. Subsequently, the formula becomes a basis for feeding a diverse group of animals that can conveniently be grouped together under a standard scientific label. There are inherent dangers involved in this kind of research. Mother Nature has never easily allowed a simple means of pigeonholing her creations. Creatures that show anatomical similarities are often found to be ethologically dissimilar. Equally, the natural diets of apparently related species are, not uncommonly, totally different. For this reason, irrespective of any other, it is always necessary to develop diets that are suitable to the animal as a representative of a particular species, not as a representative of a group. The weakness of most modern methods of wild animal nutrition is generally the failure to consider the secondary factor of feeding—enjoyment, the sheer excitement of eating! In the wild state food availability fluctuates according to the seasons, as does the preference of animals for certain kinds of food. The zoo biologist's problem is to prepare an artificial diet that corresponds to the variable wild one, without omitting to provide the correct proportions of proteins, fats, minerals, carbohydrates, vitamins and trace elements normally required.

Scientific diets are, without any shadow of doubt, a great development in the field of zoo biology, as long as they are sensibly used. To date, they have been utilized in a way that is akin to the modern methods of farming, which is quite contrary to all the basic biological laws. We are not trying to domesticate wild animals and, therefore, techniques of domestication are non-applicable. Diets that comprise nothing more than cattle nuts or special 'cake', balanced and nutritious as they may be, are very monotonous, and monotony in any form is unnatural for a wild creature. Although it is rarely possible to provide a completely natural diet for zoo animals and for the most part, nutritious substitutes have to be found, the basic principle must be that foodstuffs are provided that diverge from the natural as little as possible. A reasonable balance can be achieved by feeding the scientifically-balanced diet as an additive to the appetitive one. In this way, seasonal preferences can be catered for without neglecting the physiological needs of the specimen. Anyone who has been fortunate enough to work with wild animals will recognize the need for this, as often it is essential to be able to stimulate an appetite with some small delicacy in order to put a specimen on the road to recovery after an illness or injury. This would not be feasible if the offered gift was merely a pellet or something similarly unappetizing. Fruit eaters require fruit, meat eaters meat, insectivores insects and so on. Supplements are quite in order providing they remain supplementary.

By way of illustration let us look at the tiger. A collection that is the proud possessor of these beautiful animals cannot possibly feed them on axis deer, sambar, gaur, barasingha, nilgai, tapir, porcupine, frogs, small rodents, crocodiles, locusts and fishes, all of which would constitute their natural prey in India. Even if it were feasible to provide a comparable diet, such would not be suitable for Siberian tigers, Caspian tigers or any of the remaining variations. For example, Siberian tigers are known to fare well on shoulders of pork (wild pig can form up to 40 per cent of their natural prey), yet Bengal tigers find pork unpalatable and normally vomit any meat that is so predominantly fatty. Ideally a wide range of meats, preferably bloody, are offered to these carnivores, and coated with the appropriate vitamin and mineral additives to ensure that they are adequately nourished. Pork may be consistently offered to Siberian tigers, whilst beef could predominate the diet sheets of others. In all cases they would hopefully be offered a constant variety of choice. Variety is not enough, however: there are other considerations.

Natural selection did not equip the tiger with claws and teeth for decoration. These are the tools developed for seizing and tearing flesh. They complement the physical conformity of the creature in such a way as to make it, in terms of design, a perfect killer. This statement contains an implication that necessitates explanation. What is implied is that these highly specialized organs, organs that have taken millions of years to perfect, if neglected would become atrophied or degenerate. The suggestion has monumental ramifications. If true, it clearly indicates that unless such factors are considered, the whole concept of maintaining wild animals in captivity is pure bunk. Having made the statement, the onus is upon us to provide some positive proof. Adhering to our example of the tiger, we shall, therefore, attempt to prove this in a simple way.

American zoological parks have, possibly more than any other national organizations in the world, advanced the studies of nutritional requirements of wild animals in captivity. So well-versed are they in their understanding of these requirements that certain zoological parks have developed specific diets for felines (which include our tiger) that contain everything that the animals could possibly need nutritionally. In a typical laboratory fashion, they have got so close to the fundamentals of the problem that they can no longer distinguish the wood from the trees. As always, the danger is that if one becomes preoccupied with one aspect—that is, the nutritional requirements—one tends to forget the physiological and even the psychological requirements; yet they are, of course, synonymous. A number of these American carnivorous diets, balanced as they are nutritionally, resemble a type of pâté. Foodstuffs of this nature, prepared for specialist carnivores, quite literally deny the fact that the animals possess claws and teeth! The masticatory apparatus (jaw muscles, et cetera) not only requires exercise, but also has a profound influence over eventual skull development. There is a wealth of evidence to endorse the fact that captive carnivores fed predominantly on soft foods evidence highly abnormal skulls.

Take our magnificent tiger. He needs jaw muscles that are exceedingly powerful and such muscles need a base upon which to attach themselves. In fact his skull is not large enough to provide an adequate area for this necessary attachment. How then is it possible to produce these powerful jaws of such dreadful propensity? The skulls of wild tigers possess a feature known as the sagittal crest. This is a development of the skull that grows upwards, in the form of a ridge, providing a greater area than would normally be possible for the attachment of muscle fibres. Its growth is stimulated by the action of the jaws. As the jaws work harder, the skeletal system compensates by providing a larger area for muscle attachment, and thereby produces stronger jaws. At the same time, dealing with tougher food requires stronger neck muscles. These, too, are provided with greater areas of attachment.

Is it unreasonable to expect the reader to appreciate the implications that reflect upon all aspects of animal husbandry with every species kept with the artificial environment of the zoological collection? Those of us directly concerned with the preservation of wild animals in captivity have a direct obligation to both man and animal. If we are to have any hope of fulfilling this responsibility it is essential that those who manage and staff our collections are conversant with the basic laws of the subject.

Museum analysis of skeletal remains has often provided some of the most convincing evidence of the deformities found in captive specimens. We know, for instance, that skull abnormalities are always more extreme in the carnivorous than the herbivorous species, and the brief reference to the tiger gives an example of why this is so. However, should the reader be tempted to deduce from this that the carnivores are at greater risk in captivity than any other type of animal, it must be stressed that this is not so. Animals of the same species will often grow to different sizes, not, as might be imagined, because they represent different geographical variations of the species, but in relation to the amount of space available to them. It is a well-known fact that the growth rate of certain creatures—for example crocodiles, alligators and caimen—can be accelerated or slowed down according to the size of the area they are permitted to occupy. Possibly it is less well known that the guiding principle applies to the majority of life forms. This is why captive animals are so frequently found to be above or below the 'norm'.

While on the subject of degeneration in captivity, it is perhaps proper that we should quote one very famous example of deformity caused by unsuitable captive conditions. Earlier this century the Berlin Zoo was the proud possessor of one of the largest gorillas ever to be kept in captivity in Europe. His name was Bobby. Bobby arrived at Berlin as a 35 lb baby and grew in seven years to a giant of over 600 lb. As he grew, it became apparent that the gorilla had developed a very abnormal posture of the head. It would seem that the head was pushed up and back so that the base of the skull was in constant contact with the neck and shoulder muscles. Time was to prove that this strange posture eventually produced a deformed skull. But how did this tragedy occur?

Fortunately there is an authoritative account of the case on record. Careful analysis of Bobby's skull led to a series of conclusions as to the cause of the primary problem—the bad posture. As a baby Bobby had been cared for by a devoted team of keepers and, quite understandably, in the course of their duties, they had often carried the small ape cradled in their arms in much the same manner as they would have tended a human infant. Affectionate and well meaning as this was, it was nevertheless quite unnatural. Female gorillas do not move through the undergrowth cradling their infants in this position. Lying on his back in his keeper's arms, Bobby was forced to bend his head right back in order to look his foster parent in the face, and, when placed on the ground, his head had to be raised even more for face-to-face contact to be maintained. Young apes involved in a relationship with human beings nearly always try their hardest to look at the face of the people looking after them and Bobby was no exception. This meant that he was constantly standing in the wrong body position with his head upraised. The abnormalities of the skull that resulted were quite unpredictable because no one at that time could have been expected to foresee the problem. Nowadays young apes are usually kept in accommodation that has a raised floor so that they can look their attendants in the face without effort, and the practice of carrying them like little human babies is strongly discouraged.

Examples of this nature abound and are too numerous for us to refer to here, but the case of Bobby clearly demonstrates the tightrope walked by those concerned with biology in zoological gardens. Things that may seem important are often of little consequence and that which may appear insignificant is sometimes of critical importance.

The medical side of zoo work is another aspect with which all zoo staff should be familiar. We are not suggesting 'home doctoring'. Quite the opposite—we have witnessed far too many deaths as a result of this practice. One tragic example of 'amateur veterinary practice' involved a chimpanzee. It had to be caught and the only way of effecting its capture was to administer an anaesthetic with a dart-gun. Scorning the suggestion that a qualified veterinary practitioner should be in attendance, an over-confident management of a zoo proceeded with the operation under their own initiative. The animal was overdosed and died. The only occasion when the use of these narcotics might be considered justifiable without veterinary supervision is in an emergency, if a potentially dangerous specimen escaped from its enclosure, for instance. Otherwise all drugs should be administered only under the directives of a qualified person.

But there are aspects of animal medicine with which staff should be acquainted, and the reasons are twofold. Veterinary science is, despite recent progress, still somewhat in the dark ages, thus research constitutes a fair proportion of the zoo vet's duties. The foundations of this research are observation, and observation is one of the prime tasks of the animal attendant. The vet, therefore, depends greatly on the information available to him in the form of the specimen's history, its behaviour, whether it is captive bred or was wild caught, and the medical attention it has received during its life.

If this information is not available in the form of accurate records, the vet starts his examination blind, so to speak. As most zoos lack a resident veterinary officer and fail to provide these records, it is not uncommon for the vet to be called in to see an animal that has reached the stage where medical attention can achieve little. A vet of our aquaintance actually said, when referring to a local zoo that used his services: 'For the main part I seem to be called upon to deliver the last rites'.

If, on the other hand, the vet is able to attend to an animal satisfactorily, it is of paramount importance that the keepers are capable of following the vet's directions. Several examples of when they were not spring to mind. We have come across dozens of cases where animals on courses of drugs have shown little response to them, and in several of these the cause was not the ineffectiveness of the drugs but the fact that the keepers responsible for administering them had failed to do so. The excuses generally centred on the difficulty of administering the required dosage because of the animal's reluctance to swallow anything strange. Unfortunately many medications are strong-smelling or unpalatable and it is true it is frequently a problem to get an animal to take its medicine. Nevertheless, the keeper should report accurately the amount taken and in the instances we are familiar with they did not. In fact one case came to light because a keeper who was supposedly treating a pair of juvenile cheetah was found to have run out of tablets some days previously. He did not even offer an excuse for this neglect, simply saying that he had 'forgotten' to ask for a further supply.

The two worst cases of neglect during the aftercare of injured or sick animals that we have witnessed took place in two British parks. The first involved an ostrich that had been injured by a vicious kick from one of its companions. The central claw of the aggressor ripped a deep wound, about 8 inches long, in the bird's side, just below the wing. The wound was cleansed and stitched by the vet who was called in to attend to the casualty, and the ostrich was injected with a large dose of antibiotics to safeguard against infection. All the keepers had to do from here on was to keep the wound clean by spraying it daily with an antibiotic until it was time for the vet to remove the stitches. As the ostrich remained in its large enclosure, the only time it was possible to examine the injury was when the birds were fed, so the procedure adopted by the keepers was to clean and spray the lesion whilst the bird was eating, or at least this was what they said they were doing. About a week after the incident the writer was driving past the ostrich enclosure when he noticed that one bird seemed to be missing. He summoned the rangers and they quickly confirmed that they had not seen the injured ostrich since they had sprayed the wound the previous evening. The paddock was searched and the ostrich was found lying in some undergrowth at the far end. Even a cursory glance was sufficient to confirm that although alive it was in a bad way. It could not stand and its breathing was laboured. Closer examination revealed the ghastly truth, the wound had never been cleaned or treated with the antibiotic and it had become fly-blown. Now it was a grisly, suppurating mess beyond hope of repair. The bird's death can only be attributed to neglect.

The neglect of a cheetah in another park can be put down to sheer stupidity. The animal broke its leg, a catastrophe for any animal, but particularly so for these fleet-footed athletes; but the vet subsequently set the bones most skilfully. At this park the cheetahs slept together in a wooden shelter, the heat from their collective bodies keeping the ambient temperature at a reasonable level during even the coldest night, but obviously it was not possible to place the injured animal in with the others. For some time, at least until the removal of the plaster, it was necessary to keep it in solitary confinement, and this is exactly what the park management did. What they failed to do was to provide it with a secure, warm place. They just moved it into a shed similar to the one its companions slept in, and left it. Without heating one might expect problems with any animal which has just broken its leg, an animal that is in a state of shock, but this was mid-winter. The cheetah froze to death on the first night.

The health of wild animals in captivity does not rest on diagnosis and treatment alone. In fact, careful management is intended to preclude the need for more drastic measures. The emphasis is on prevention being better than cure.

Not long ago nearly every wild animal kept in a zoo was brought there from the wild. Even now a large proportion are captured and brought to British zoos from their country of origin and this necessitates adequate quarantine, not only to meet the Ministry requirements to keep out diseases that would affect our domestic pets and farm stock, but also to protect the public from diseases communicable to man, such as rabies, psittacosis, salmonellosis, ringworm and so on. Equally important to the zoo owner is the quarantine to prevent these new introductions from carrying diseases that would infect the stock already in the zoo. It also gives the opportunity to the veterinary staff to examine thoroughly and diagnose and treat animals carrying diseases in a chronic form that are likely to flare up because of the added stress of captivity and acclimatization. For example, most wild caught animals carry a rather large burden of internal parasites such as worms, fluke and larvae of various kinds. They may also have blood parasites which would be highly dangerous to other zoo stock and domestic animals and indeed quite a number of infectious and contagious diseases can lie dormant in the imported animals until some unacceptable stress precipitates an active infection. We are thinking of not only the stress of captivity but also dietary imbalance, an increasing worm burden through animals being kept on small areas of land causing a build-up of infection, and social problems such as the stress of overcrowded quarters, repeated attacks by more dominant animals and climatic changes which are not cushioned by comfortable warm sleeping quarters.

Not only are most animals quarantined in the country of origin close to the point of capture for about a month, but they also have to spend at least six months in quarantine in the United Kingdom on arrival. It is during this time that many examinations and tests can be made and treatments carried out to reduce the worm burden and clear the animal of any latent diseases. It is while in quarantine that the animal becomes acclimatized not only to change of temperature and humidity but also to a change of diet and living conditions and adapted to the proximity of man. It is said by most zoo authorities that if the animals, and in particular birds, survive the first year they will rarely give trouble afterwards. This is not to say that deaths during quarantine are high for most mammals, though they are certainly higher than they would otherwise be.

It was found, for example, that certain types of penguin rarely survived in captivity for very long and death in most cases was due to aspergillosis, a fungus infection of the lungs. It seems that a few birds have small foci of infection in their lungs when they are caught in the Antarctic. Because of overcrowding on the ship during their transport to the northern hemisphere and the climatic changes, the disease not only spreads among the birds, but becomes a serious problem in many. It is a difficult disease to treat because although there is a therapeutic substance that will kill the fungus, this agent is almost insoluble and so difficult to apply. If given by mouth very little of it is absorbed into the blood-stream and it cannot be given by injection. And so, on arrival in the United Kingdom, many of the penguins are suffering from an incurable disease that tends to get gradually worse.

Tuberculosis in monkeys is another such problem, and salmonellosis in a number of species, particularly terrapins, tortoises and various species of birds.

Fortunately the most prevalent problem, namely internal parasites, can be treated, and zoos have developed a routine system of faeces examination for the eggs of the parasites and regular treatment with the correct anthelmintic. There is no doubt that reducing the worm burden not only prevents the animal from becoming debilitated but also allows digestion and absorption of food material to proceed more normally.

Unfortunately, we have to return to our point that the success or failure of parasite control, as with most aspects of wildlife management, rests solely on the quality of the staff and the accuracy of the zoo records. Worming is routine procedure in some collections, but many cannot recall with any degree of certainty which animals were wormed or when. We watched an autopsy on a giraffe in the early seventies. As the vet's scalpel sliced through the animal's intestines thousands of ascarid worms issued forth. If the group had been treated regularly this could not have happened. One park in England lost several giraffes from lung fluke, another a whole herd of the rare European bison from a variety of parasitic infestations. Control of these dangerous parasites depends totally on the zoo management ensuring that regular faeces tests are performed and that regular worming is standard procedure, and on the keeper making certain that each individual animal receives the required dosage. Too often have we seen an animal attendant deposit a sackful of worming pellets in a trough in the vain hope that the herd will consume equal amounts of the anthelmintic. What actually happens is that the dominant animals eat the majority whilst the subordinates get little or nothing. If a worm problem exists within the group the exercise achieves nothing, for the infected specimens that are not treated will quickly re-infect their comrades.

The zoological park must therefore determine to control these parasites and an important line of attack is to ensure that the animal quarters are cleaned regularly and disinfected, that food is placed in proper receptacles—never on the floor, as this is the best way for stock to re-ingest larvae and eggs—and that alternative enclosures and isolation areas are always available in the event of a flare-up of infection. The last point is but rarely considered in most of our zoological parks, which is foolhardy because, if such provision is made, it can mean the difference between one or two animals falling sick and a full-scale epidemic.

As a general rule, the cleaner the enclosures and animal houses, the healthier will be the stock. Yet, having stressed the value of hygiene, a word must be said on the extent to which it is taken. The aim is to achieve cleanliness, not surgical sterility, or asepsis, because there are no aseptic conditions in the natural world. Zoos must not confuse their exhibits with laboratories, although they should endeavour to maintain their animals in hygienic conditions. It is all a question of balance. Over-enthusiastic and excessive cleaning is often as bad as neglect. Disinfectants are spilt into feed dishes and water troughs, floors are left wet, creating a slippery, dangerous surface, or fresh straw is laid on them before they are dry, leaving a damp bed for the animals to sleep on. Moreover, environments that are aseptic are quite literally unbiological. This is contrary to all zoological objectives and can ultimately create manifold problems, particularly behavioural ones.

Humans associate disease with smell and when involved in routine cleaning operations, try extremely hard to remove both. In the process they seem to forget that for many animals the olfactory sense is as important as sight. By removing the animal's own scent from its quarters, it is quite possible to cause stress and insecurity. Many species use scent-marking as a means of identification and ownership. The male rabbit marks the youngsters of his own group with a secretion from a gland situated under his chin. Similarly, young tree shrews and infants of certain species of mongoose are marked by the closely-related adults in their group. Once the young are marked, there is no danger of accidents caused by wrong identification: a strange infant would be driven off, but if the adult animal is assured by a cursory sniff that the little one belongs, all is well. The domestic dog is another animal that displays behaviour influenced by scent and many an ingenious zoo director has turned this to his advantage when in need of a foster mother. Orphaned babies like lions and tigers can be successfully reared by a bitch that has just given birth to puppies, but before she will accept her strange child it is necessary to impregnate its coat with the smell of her own young. Urine from one of the litter rubbed over the feline baby is usually sufficient to deceive her; if it smells like a dog, it must be a dog! Coversely, if, say, a leopard gives birth and her young are continually handled by the human attendants, she will often reject them because they no longer smell of leopard.

Apart from scent-marking con-specifics, territory is also marked and in a variety of ways. Claw marks on trees are almost certainly territorial signs and certain species of deer slash bushes and shrubs until they lie in shreds; this too is a type of territorial 'sign-posting'. There are in fact a number of kinds of territorial signals, but we will consider the commonest form, that of scent.

Scent may be deposited by the sophisticated development of special secretory glands. These special glands are found throughout the animal kingdom and on just about every area of the body. There are anal glands, caudal glands, glands on legs, feet and even between toes. Secretory glands exist by the eye, on the ears, over the face and head, on the chest and tail, and some species have a secretion in the saliva solely for the purpose of marking. Best known of all the territorial marking techniques is that in which faeces and urine are deposited at selected sites as a statement of ownership. Everyone will be familiar with such behaviour in the domestic dog. If taken to an area that is a favourite haunt of dog owners, a male dog will cock his leg against a multitude of objects well in excess of his need to relieve

13 The fate of the elephant?

a. Several thousand elephant skulls confiscated from poachers at Tsavo National Park, Kenya

b. Elephants at a wildlife park in England

14 Siberian tigers at Marwell Zoological Park: two mothers, father and five cubs. Note swimming-pool at back

15 Bliss in a natural environment: a hippo wallows among water hyacinths, Gorongoza National Park, Mozambique

16a. Deprived environment for a curious-minded species

b. Sexual arousal can escalate to abnormal levels

c. A Sun bear plays the 'titbit game' to reduce boredom

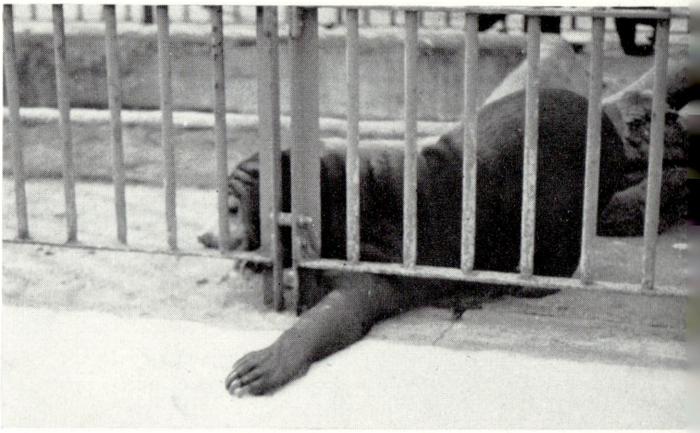

d. Do you find me interesting, funny or pathetic?

17 A family of Eurasian otters bred at the Otter Trust, Bungay, Suffolk

18 Young European Eagle owls bred at the Norfolk Wildlife Park, Great Witchingha

19a. Baby orang-utan born at Twycross Zoo Park
b. Infant gorilla at Jersey Wildlife Preservation Trust

20 Dosing Polar bears during an anthrax epidemic

the bladder. He is simply placing his mark over that of other dogs and temporarily staking a claim to the area. In his own garden the dog is not so obsessive about marking; the scent lasts for 10 to 14 days and requires only periodic freshening up, so for as long as his 'territory' retains his characteristic smell, the animal will feel secure and content. Introduce the scent of a strange dog, however, and the resident animal will spray every contaminated tree and bush until all trace of the foreigner is completely erased.

Urine marking appears to be the rule of all male members of the *Canidae* (dog family) although the females of some species have been observed using it, whilst others only mark when in oestrus. A gland is present on the dorsal surface of the tail in many members of the group that use an underground den, and this possibly serves as a secondary means of transferring scent whilst entering or leaving the den. Here we have a case where two types of marking techniques are employed, a factor that is not uncommon.

Field observations of the brown bear have shown that these animals normally select a number of trees to mark and they use a variety of methods to do so. They certainly spray the chosen tree with urine and also disfigure the trunk with deep claw marks. In addition, the bears have neck glands which are rubbed against their signpost, placing yet another scent on it. Rubbing the neck glands against the tree is often accompanied by scratching or chewing and, not surprisingly, there is a strong indication that the chewing represents a fourth type of marking, this time with saliva.

The significance of scent marking to the animal can be reasonably well understood if we consider the individual specimen's response to the two distinct kinds of scent that it will normally encounter: namely, its own mark or that of a foreigner.

An animal's response to its own mark, or in social species, that of any number of its group, will be a noticeable reduction in anxiety accompanied by obvious signs of increased confidence and contentment. By impregnating the whole living space with its own scent the animal warns others of its own species that this territory is occupied. Thus, by excluding con-specifics without the need for physical contact, the animal creates for itself a secure environment that can be considered a true home.

Conversely, if an animal encounters the scent of a foreigner, the emotional effect will be distinctly different, although its ultimate expression will depend on whether the responder is on home ground or not. If on its own territory the animal will show a marked increase of aggression and readiness to fight. If trespassing on another's home area, there will be a decrease in confidence and a strong urge to flee. A third type of behaviour can occur if the marker and the responder are of opposite sex—that of sexual arousal and mating.

Scent marking can, therefore, be neither ignored nor its significance to the animal underestimated by the technician responsible for the daily care of wild animals. Aseptic conditions would presume to remove all scent marks daily. This means that the animal continually 'loses the ownership' of its territory and is forced to repeatedly stake its claim by remarking its cage or enclosure. It is but a short step from this point to an abnormal increase in the action culminating in stereotyped scent-marking behaviour, or marking that no longer serves a sensible purpose.

Quite obviously, it is essential that the animal attendant is well acquainted with these phenomena if complex behaviour problems are to be avoided. He must always bear in mind that, whilst the animal's zoo environment must be kept clean and free of infectious material, the creature's identity should not be removed along with its dirty bedding. Quite apart from avoiding constantly removing the scent marks from the cage, it is just as important to ensue that adequate provision has been made for the depositing of these marks. Dr Desmond Morris cites the example of a colony of blackbuck in his paper, 'The Response of Animals to a Restricted Environment' (1964).

Blackbuck, like many deer and antelopes, have a gland situated by the eye, which produces a strong-smelling, oily substance that is rubbed on to twigs and branches. The animals in question were living in a large, natural colony group, but had been restricted to winter quarters in which there were no low branches or bushes. In the absence of suitable marking points, the animals resorted to using the tips of one another's horns as scent posts—a practice that involved a considerable risk to their sight. In this case the behaviour of the marker had not been properly considered and as a result the animals' welfare was severely jeopardized.

These, then, are some of the facts that the animal technician must understand before embarking on the seemingly simple job of cleaning his charge's zoo environment.

Occasionally, in spite of all the care and attention to detail, disaster strikes, and one such example struck Chester Zoo some years ago. It began at noon on Wednesday 18th March. A female African elephant died. She had been repelling the attempts of two male African elephants to mate with her all night and the keeper had come to her rescue in the morning by shutting her off alone. She was exhausted, and the keeper, becoming alarmed at her weakening condition, called the Director, who in turn telephoned the vet, who hurried to the zoo to find her dead, apparently from shock. She had not been in full health since acquired two years earlier. An infected tusk socket had cleared up after several injections, but three months earlier she had fallen, cracking a tusk, which dropped out, leaving a deep, infected, cavity, which she repacked with soil every time it was cleaned out. The pain and swelling subsided but it discharged occasionally in spite of antibiotic treatment.

In such a large community of animals, deaths occur. The life-span of most animals is far shorter than man's and so death is more common than in a comparable village of a thousand inhabitants. It is the veterinarian's job to establish the cause. So it was not extraordinary to have three racoons and one grison die the following Sunday. Two racoons had signs of acute pneumonia but the other showed very little that was abnormal. The grison had a diseased liver. There seemed nothing in common with these deaths and yet the vet had a nagging feeling that something serious was wrong.

Diagnosis of disease in animals is not easy. They do not answer questions. Wild animals present a greater problem. It is difficult or impossible to handle them, to feel their pulse, take their temperature, or listen to the lungs and heart, because they are either dangerous, or frightened, or both. They can hide their symptoms, appearing normal, to the point of death. A post-mortem examination often shows such extensive pathological changes that one wonders how they lived so long. Yet the reverse can be seen, and no reason for death can be determined.

Blood smears from the three racoons and one from the grison were made. They were positive for anthrax, one of the most deadly of all diseases.

Herbivorous animals are the most susceptible and though it was 10 p.m., the vet hurried to the Director with the bad news. Of all the herbivorous animals, only the elephants were shut indoors, and, as they are so valuable, they were examined. They had eaten and appeared normal and a little surprised at a visit so late at night. Without help the vet could not inject penicillin, which was the best treatment for this disease. All seemed quiet in the zoo—the lull before the storm.

The following Monday was a day never to be forgotten. Two large elephants, weighing three tons each, lay dead, another smaller one was ill with a temperature of 105°F. Two more were in mortal danger, as were most of the animals in the zoo. The bodies had to be removed and burnt—the law requires this. The others were put in isolation and all had to be given penicillin daily.

Each species is a special problem. For instance, tablets can be given to animals with simple stomachs, for they can be absorbed rapidly; but in those species with large stomachs and complicated digestive systems—in other words, the herbivores—the tablets could lie for days and not be absorbed. Hence injection was the only way.

Injecting an elephant is not an enviable task. Penicillin was needed quickly and in large amounts. Certain things became clear; it would be impossible to inject the herds of deer, giraffe, gnu and zebra that roamed the paddocks. At least not without great risk of causing so much excitement and fear that injury must occur.

From where had the bacillus come? Where could it have spread? It is found only in blood and meat and unsterilized animal products. It can remain dormant in soil for years.

Each section of the zoo has its own supply of hay and grain. There was a good chance that it had spread no further than the central area. Three racoons and one grison had died and so the whole of the butcher's area, which included the fish supplies, was suspect. This meant that the carnivorous animals and birds and the fish-eating mammals, such as the polar bears and sea-lions, were at risk.

Anthrax has never been known to affect carnivorous birds, and no case has been reported in any cat species. So the plan was made. The vet would inject the elephants with immense doses daily, and also inject the polar bears and sea-lions. All the other carnivorous animals would get tablets according to weight, in their food each day. All the animals were carefully watched and a clinical round was made each morning.

Of the three elephants left alive, two were robust and healthy and daily injections with 15 million units of penicillin presented no problem. The skin behind the ear is thin enough to allow penetration with a hypodermic needle. Only by standing on tiptoe could the site be reached, and while doing so it was unnerving to have a sensitive but rather strong trunk examine one's hands, syringe, legs, body. With only the keeper present, a pocketful of dates as a reward and plenty of time, it was accomplished eventually.

The third elephant had a temperature and did not eat. She was dull, her head and trunk hanging dejectedly, eyes running and the conjunctiva bluish and swollen. Her heart could be heard clearly thumping in her chest. It was no problem to inject her twice daily.

Meanwhile, on Tuesday, two racoons were found dead and we discovered that they had been washing the penicillin tablets out of the meat in their drinking water. Blood examination was positive. As three other racoons were ill it was decided to inject each day all the six remaining animals. It was not easy and a keeper got bitten. The largest polar bear was more of a problem: they are immensely strong and dangerous, and this one could not be trapped in a crush cage. Injecting was impossible and as he carefully removed the penicillin tablets from the fish or discarded them altogether, it was difficult. Finally he was enticed to open his mouth for a tempting fish and then penicillin suspension was squirted down his throat with a syringe. Happily, he and the racoons all recovered.

A badger died on Wednesday, also positive, but there was more hope now. The sick elephant seemed a little better; she drank three gallons of milk and bread and slept a lot. On Thursday she was further improved, warmer, eyes more open and attempting to eat hay, though the conjunctiva was still blue. Then tragedy—one of the larger elephants attacked and knocked her down. She got up and they were separated. But on Friday she was dead—yet her blood was negative—so near her recovery.

She was the last to die. Altogether, since that horrific Monday morning, only two racoons, one badger and the sick elephant died—a tribute to the efficiency of penicillin. Of course, treatment continued for another full week. And the quarantine—a regulation imposed by the Ministry of Agriculture—was lifted ten days after that terrible Monday. The elephant carcases were burnt with two powerful oil burners suitable for a central-heating furnace. And what could have taken two or three weeks using traditional wood and coal was accomplished in four days. One man contracted the disease, fortunately as a skin sore. It was diagnosed early and hospital treatment in strict isolation effected a rapid cure. And so ended the most dramatic outbreak of anthrax in a major zoo. The only other case in elephants in England occurred in 1926–27. But in Africa, Burma and India it is relatively common. Mentioned by Homer, described accurately by Virgil, it is still a threat.

This particular zoo had a small hospital, a clinic and veterinary supervision. Every animal that died was post-mortemed, and the disease was picked up quickly and treatment commenced, otherwise many more animals might well have died.

The veterinary surgeon in attendance not only makes routine inspections of the zoo and examines those animals that the keepers are worried about or that he himself detects as appearing abnormal, but also takes the necessary samples and examines them in the laboratory attached to the zoo. If the disease is serious, the animal is isolated but if not, experience has shown that they are far better left amongst their companions, firstly because it is an added stress to move them to fresh surroundings when they are feeling ill, and secondly it is difficult to return such an animal once recovered to his original companions, for while he is away they have readapted themselves to a new dominance hierarchy in which there is no room for the animal on its return.

The minimum requirements for any zoo, therefore, must include a small clinic and equipment for catching and handling the animals for examination and treatment. It must also have one or two isolation rooms, not only for its own safety, but also to protect the public and to keep the animal out of public view. Some of the larger zoos are equipped with hospitals, but many experts believe that it is better to have a hospital-type room in each section of the zoo where the animal can be looked after by its own keeper and kept within smell and call of its own companions. Capturing equipment is essential. In the past, before the arrival of the tranquillizing gun, zoos equipped themselves with various compartments and trap-doors behind each group of sleeping quarters, into which the sick animal could be coaxed or driven. It would be in sight of its own companions and yet protected from their aggression which sometimes occurs when a dominant animal is taken ill. From these small compartments the animal could easily be squeezed into a crush cage with collapsible ends and walls in which the animal could be firmly and safely held while it was examined, samples taken and it was treated, or, if necessary, anaesthetized.

For many species this is still the best procedure and zoos are returning to this system now that the glamour of the tranquillizing gun has faded. Certainly the gun is a very valuable piece of equipment for quietening and translocating frightened or aggressive animals. It is less valuable for sick animals for obvious reasons. The depressing action of a tranquillizer may make the animal much worse. In any case the dart that is fired from these guns is fairly heavy and though little problem to a large animal, can cause a fair amount of injury to a small species. The dart must be shot into a large mass of muscle and this is not easy with small animals such as a monkey high up on a tree. Far better to trap the animal by shutting it in when it goes to rest at night.

We have touched on only a few of the problems of animal management, but from this maze of considerations it is surely clear that veterinary and research facilities must be available in all collections, and staff must be trained accordingly. For so long as our zoos remain under-capitalized and persist in employing unskilled cheap labour, instead of training suitable people for the job, the zoo is heading nowhere. Preservation is truly barred.

Chapter 8

# The best of our wildlife collections

The bulk of this book has been critical—critical of systems of capture and transportation, critical of the animal-dealer—but primarily our attack has been centred on the British wildlife collection, its poor standards and general lack of objectives. We have stated that the keeping of wild animals in captivity can only be justified on the grounds of research, education and conservation and then proceeded to say that, to date, the achievements in any of these fields is so negligible as to be virtually insignificant. Yet in the introduction we were bold enough to insist that the survival of the zoo is essential, that it has a significant role to play in modern society, that for many species the zoo offers the only chance of continued existence, but since then we have said little that substantiates this optimistic defence of the institution. Indeed, if the reader were to assume that we were just totally anti-zoo, one could hardly blame him. Nevertheless, the assumption would be incorrect. The views expressed in this book are our own; many will surely disagree with them, but having expressed our opinion on what we consider the shame of the British collection, we are clearly obliged to offer some form of solution. To this end it seems reasonable to focus attention on the handful of British collections that are already doing an admirable job and can be considered as signposts to the future. This should enable the reader to go and look for himself and with luck an entirely new concept of the zoo's potential will open up before him.

Such a selection must be biased, at least biased in favour of those who have always managed to command the respect and admiration of the authors. Therefore, one must understand that the collections mentioned are not perfect, there is no such thing as a perfect zoo, but in our estimation they have shown themselves to be completely committed to the welfare of wildlife and to have the ability to achieve a great deal in its interest. There are obvious omissions, the most obvious being the safari park and the large Society mainstream zoo; this intentional disregard for their existence demands explanation. Safari parks have been excluded on the grounds of our own personal experience of them, which has been far from favourable, and the seeming lack of positive breeding results or research. They will understandably leap to their defence with success stories like the breeding of cheetah, but chance breeding cannot constitute a defence. On the safari-park organizers' own admission they are created to make money and that, despite other allusions to moral objectives, is their basic function. We have never denied their potential, we have merely damned the fact that the potential seems to remain uninvestigated. So in our view the safari park has spread through the civilized world not like a brilliant concept of wildlife management, but like an insidious cancer that is antagonistic to the whole principle of wildlife preservation.

The gigantic Society zoo, the mainstream collections like London, Chester, Dublin, Edinburgh and their smaller counterparts like Bristol and Glasgow, have been omitted because we reject the concept, not because we suggest neglect of the animals or crude commercial exploitation. This style of collection, which has produced the image of the zoo that is the standard image, the rows of animal cages, the large animal houses, the uniformed keepers, was dreamt up 150 years ago and its function, its objectives, even its mode of daily operation are as relevant to modern-day zoological thought as the politics of 1826 are to present-day political aims. To be honest, we feel that the only major collection of this type that shows signs of breaking through the barrier created by an assembly of aged and dictatorial councillors, that shows signs of remaining up-to-date, is Chester Zoo in the north of England. Still, even then, we reject the initial concept. No one can hope to be an expert on all aspects of natural history yet this is what the director of a large general collection endeavours to be if the project is to be of legitimate value. It is of course a virtual impossibility to be such an expert, and the present conditions found in most mainstream zoos seem to bear witness to this—none seem particularly relevant to the current crisis facing man or wildlife.

What we are looking for is specialization; wildlife collections that are aware of the fact that individual groups require distinct forms of accommodation and styles of husbandry, and attempt to satisfy these requirements. There are several of this kind of collection to be found in Britain, yet as a general rule their presence is missed by the average citizen. Therefore, we shall cite a few examples of the cream of our British collections, chosen not because they are necessarily the most attractive or grand places to visit, but because they have achieved more than most and will doubtless continue to do so.

Since 1970 several new collections have opened to the public, but only two of them are particularly worthy of mention. Coincidentally, the men behind them are both, academically at least, unlikely individuals so far as the conventional image of a zoo director is concerned, one being a businessman, the other a gambler.

The arrival of John Knowles on the zoo scene in the late 'sixties was cause for much speculation amongst zoo personnel the length and breadth of Britain, for he was a talented and, therefore, wealthy businessman who was planning to start his own zoo. Not that that in itself was particularly unusual, many men with similar backgrounds had opened zoos and managed to make substantial profits from them, but Knowles was different because he talked of establishing a collection of rare species solely for the purpose of research and preservation. The commercial organizations chuckled amongst themselves and predicted his financial downfall, whilst the more academic institutions tended to doubt his sincerity.

The opening of Marwell Zoological Park in Hampshire in 1972 finally dispelled all doubts as to his solvency and his sincerity, for it established the most important major zoological collection to open in Britain for many years. It is not another 'conventional' zoo. On the surface it is an awe-inspiring collection of some of the rarest and most beautiful animals in the world, many of which are seldom seen in zoological collections. But behind the mere visual impact is its potential, for it portends to become one of the leading zoos in Europe and a major contributor to wildlife preservation and research. Of prime influence on John Knowles were undoubtedly some of the American zoological collections and, of these, it is probable that Roland Lindemann's Catskill Game Farm had the most profound effect. Catskill, a private zoo on a 900-acre site, is world-famous for its breeding of Przewalski's wild horse, onager, mountain zebra, scimitar-horned oryx and many of the other species of large ungulates. Although Lindemann is well known as a dealer, it should be pointed out that he is of a different mould to the general run—first and foremost he is a breeder and passionate conservationist. This influence can be clearly seen in the presentation of the Marwell collection and also in the style of some of the park's publications and this, plus John Knowles' own unique stamp of individuality, provides a freshness that has long been needed in the British zoological park.

Set in attractive Hampshire parkland that surrounds a magnificent old manor house, the collection has been arranged on a zoogeographical basis. This means that as the visitor proceeds through the collection he moves from Africa to Eurasia to America and Australia. From a simple educational viewpoint this can hardly be bettered, for the average visitor is notoriously ignorant of many species' places of origin and this is a subtle and stimulating way to teach them. Also it is aesthetically more pleasing to see species from the same continent grouped together, so it is surprising that more zoological parks have not developed this ingenious style in preference to the mindless groupings so often seen.

At the time of writing, 150 acres have been developed and are open to the public, displaying 53 species of mammals and 28 species of birds. Although this is described as a mixed general collection, it should be emphasized that it centres around a large and very specialized collection of big cats and ungulates. The specialist interested in comparative zoology can wallow in the glorious excesses of Marwell. Here one does not just see zebra, but Chapman's zebra (2 born in 1974), Grevy's zebra (4 born in 1974) and Hartmann's mountain zebra, plus the extremely rare Przewalski's wild horse (5 born and reared in 1974), the Persian onager and kulan. Also to be seen here are the dramatically beautiful scimitar-horned oryx, of which Marwell bred no less than seven in 1974 and which have been adopted as the park symbol, the equally dramatic sable antelope, nyala (the first to breed in this country for over 100 years), nilgai and the unusual goral. Unlike the more conventional collections, these exotic ungulates are not normally kept in two's and three's, but where-ever possible in large herds, and this results in not only the most spectacular display of herbivores in the United Kingdom, it also promises to be the most viable in terms of breeding.

The deer family are also well represented and include the following species: Père David's deer, axis deer, hog deer, Formosan sika and barasingha.

The Siberian tiger and its smaller relative the Sumatran tiger are two more endangered species to be seen at Marwell. Neither species is commonly kept in captivity and it is therefore encouraging to see that five Siberian cubs were reared in the collection in 1974. In the same year Marwell successfully bred the jaguar, leopard and the northern lynx. As yet the cheetahs have not reproduced, but it can be expected that Whipsnade's success with this difficult species will be duplicated here eventually.

Although essentially a modern, progressive collection, Marwell is not an architectural zoological park; the majority of its animal accommodation is of a simple 'cattle house' style, pleasantly faced with timber. Considering the species kept, this is probably a very wise decision as the buildings are unobtrusive and highly functional, whereas vast concrete structures would blight the landscape and offer few serious advantages over the existing system. It also means that any future redevelopment will not constitute a major headache, most of the houses being simple to dismantle and move or replace. But the caging is by no means unimaginative. For example, the leopards have access to a tree and can usually be seen high up in the branches. This has meant the construction of an extremely high cage with a roof on top to safeguard against escape, but it has created the effect that the animals are not restricted in any way. The Marwell leopard cages are certainly the best in any British zoo.

We have consistently maintained that the distinction between a good collection and one that is less than satisfactory depends almost entirely on the philosophy of the park's management. This being so, we asked John Knowles for his opinion of the zoo's role in this age and, in particular, what he saw as Marwell's purpose. He replied as follows:

> Until recently it never occurred to man that there should be any other reason for keeping wild animals alive in captivity other than providing a peep-show. It is now abundantly clear to anyone who studies the growth of human population all over the world, and the growing desire of this expanding population, for what are called 'higher living standards', that the prospects for survival in the wild, in species other than those which, like rats, can live advantageously side by side with Man, is gloomy at the best.
>
> Even the game reserves, where they have been set up and effectively policed, are vulnerable because of Man's numerical growth. I therefore feel that we have now reached the point where every possible species should be established in satisfactory breeding groups in one or more well-managed collections, and the role of a modern zoo must be that of a latter-day Ark. This may ensure the survival of a species and one can think of quite a number of species already that would not be in existence today were it not for zoos, and if by any chance my view of the future is too gloomy, it would still be advantageous to have insurance stocks of wild animals in captivity and every effort should be made to learn from these animals as much as possible of their needs, so that this can be applied to assisting the wild populations.

Education, both in the formal and informal sense, in the natural world, and the need for conservation, is an important role for a zoo, as the future of wildlife eventually depends on human beings, and so a zoo can usefully be used to promote an awareness and a concern which cannot be stimulated in any other way than by actually seeing a living animal.

A zoo can only survive economically if people will pay to come and see the animals, though in a really enlightened world this should not be the only economic life-line. There is no doubt in my mind that there are conflicts between exhibiting animals and breeding them. This could be overcome if zoo visitors were more appreciative of the value of large groups of animals breeding, and understanding what the operation is really all about.

Marwell's story is just beginning. As yet it is still a very young zoological park, but it has already achieved far more than most of the large collections in the country. In the interests of breeding endangered species, reciprocal arrangements have been established with several genuinely preservation-orientated collections abroad, and in this country similar agreements have been reached with the few collections that are similarly inclined—the Zoological Society of London, the Jersey Wildlife Preservation Trust and John Aspinall's collection at Howletts in Kent.

John Aspinall's is the second collection we wish to mention, or to be more accurate we should say his two collections. Mr Aspinall made his fortune from gambling, a fact that bothers Mr Aspinall not at all, but a fact that his critics delight in harping upon. However, gambling gains were never put to a better purpose than the upkeep of this remarkable collection. It was started at Howletts in 1957 and has been maintained ever since despite the fact that he was unable to open it to the public in order to cover some of the running expenses. Today the park is only open to visitors between May and October.

The central feature of the collection are the gorillas, the largest group in Europe, and the tigers, the largest breeding colony in captivity in the world (14 cubs were born in 1974). There are similarities between the collections of John Aspinall and John Knowles. For example, they are both interested in the cat tribe and have, in addition to tigers, species like Clouded leopard, serval, leopard and cheetah. Both have exceptional collections of the rarer antelopes, and both have achieved far more by way of breeding than most of their 'professional' zoologist colleagues. Ninety per cent of the animals kept at Howletts have bred in the park, an achievement few zoos in the world could match, and many of these births are remarkable by any standards. The first Clouded leopards to be bred in Britain were born here and the animals have continued to breed regularly since. As Aspinall also has the only specimens of the rare Snow leopard in the country it is to be hoped that he will follow his success with the Clouded leopards by breeding the Snow.* All the breeding results here seem staggering. As we mentioned in an earlier chapter, a life-long ambition of John Aspinall's was realized when the first gorilla was born in April 1975, but it didn't stop there. Later the same month another was born and a month later a third; the following year saw the birth of five gorillas at Howletts. If every collection keeping these apes did half as well, the gorilla's future would be assured. The question is, how does he do it?

John Aspinall would have little doubt as to the reasons for his success, for he places great store in personal relationships with his animals. He plays with his tigers and gorillas, is allowed to enter the territory of his wolves (a pack numbering over 30 animals), where he is welcomed like the lead wolf, and he takes his Clouded leopards for walks across the estate. He also has a loyal team of staff who pursue the policy of making friends with their charges despite the fact that most zoo authorities feel that personal contact between animal and keeper is undesirable, the main argument being that there is a danger that the animal will transfer its attention from members of its own species and thus fail to breed.

* At the time of going to press we heard that the Snow leopards had bred—Sept. '77.

As John Aspinall's animals outbreed them all, one must consider the theory to be, at the very least, suspect.

Still we do feel that there are other considerations that are equally important in the Howlett's success study: intelligent grouping for one. (Most zoos are content to show the odd pair rather than a biologically representative group.) The consolidation of a devoted and intelligent staff, and spacious, well-considered enclosures and cages are others.

It seems impossible to say enough in praise of this extravagant personality, but if proof were still needed of the man's amazing abilities with wild animals, it was surely provided when he opened his second collection at Port Lympne estate, overlooking the Romney marshes near Hythe in Kent. This park was opened to the public in 1976 and by the end of the year 75 per cent of the species in the collection were breeding.

First impressions of the Twycross Zoo Park might give one the idea that it is another general mixed collection run along conventional zoo lines. In fact nothing could be further from the truth, for this is an exceptional zoo in many ways. The zoo was started in 1961 by two ladies, Miss Badham and Miss Evans, when their private collection of animals grew too large to keep anywhere other than in a zoo. Since then they have built up one of the most successful zoos in the country. Although the species to be seen here range from elephants and giraffes down in size to a variety of small mammals, birds and reptiles, the highlight of a visit to Twycross are the monkeys and apes. Here you can see gorillas, orang-utans and chimpanzees, six species of gibbon, the largest collection of this group in the country, the rare douc langur and spectacled langur, proboscis monkeys and black and white colobus. The rarest monkeys in the zoo are the weird-looking red uakari monkeys from South America and the collection of South American species is the most comprehensive in Britain. Although to date the gorilla has failed to breed, their successes outrank their failures.

In 1972 they bred the Indian smooth otter, the first time this species has bred in captivity in the Western world, and giraffes, tigers and Brazilian tapirs have also reproduced; but it is for their primate breeding that they are deservedly world-famous. Orang-utans (living in one of the finest ape buildings in the United Kingdom), chimpanzees, proboscis monkeys, langurs, colobus, woolly monkeys, silvery marmosets, and lar, agile and pileated gibbons—all have been bred at this remarkable zoo. The secret of their achievements seems to be entirely due to the two charming ladies who founded the collection. They have an astonishing degree of understanding when dealing with most animals, but with primates the depth of understanding reaches uncanny proportions.

We remember meeting Miss Badham in the winter of 1973 as she returned from visiting an animal-dealer's. In the back of her car was a box containing three of the most pathetic, emaciated and miserable little monkeys we have ever chanced to see. She had purchased them from the dealer because she was certain they would die if they remained there. The monkeys were Howlers, a South American species with a reputation for being decidedly difficult to keep in captivity under normal circumstances; but these specimens were young, in fact so young that they should never have been imported, and very sick-looking indeed. A callous view of the situation might have been that Miss Badham had, in her act of kindness, simply thrown her money away, for the monkeys could hardly be expected to survive for long. However, contrary to our pessimistic forecast, all three monkeys survived, thanks entirely to the loving care and attention given them after their arrival at Twycross. And, in the summer of 1976 Miss Badham proudly pointed them out to us, playing tag in their grassy outdoor cage, quite unrecognizable as the forlorn bundles we had seen three years before. We doubt that any other zoo in the country could have saved them. Since 1972 the zoo has been a registered charity, with all profits being ploughed back into it for the benefit of the animals.

European species are less often seen in our zoos than are the tropical species, but at Great Witchingham near Norwich one finds a park with the largest collection of European animals in the world and a splendid record of breeding successes to go with it, for it actually breeds more animals than it takes from the wild. The Norfolk Wildlife Park was started in 1961 by the well-known author, broadcaster and naturalist, Phillip Wayre, solely to encourage captive breeding and the preservation of wildlife, and since then he has established so many breeding records it is difficult to know where to begin: British brown hare, Canadian otter, racoon dog, European suslik (ground squirrel), European beaver, spur-winged plover, each at the time believed to be the first successful breeding in captivity in Britain. The European wild cat was bred here in 1970, the first time the species had bred in captivity in this country for 86 years, and in the same year two female common otters produced litters, the first time in captivity in Britain for 90 years. By 1972 the common otters had bred for the fourth time, one cub born and reared being the second generation bred in the park, the first time this has been achieved in the world. The list seems endless, but Phillip Wayre is not a man to simply add up his successes; like many of us, he sees captive breeding as only part of the task of captive preservation, and actively endeavours to re-introduce park-bred animals to their natural environment. Since 1962 over 60 European eagle owls have been bred at the Park and of these 28 have been presented to the Swedish and German Conservation Authorities to help their schemes to re-introduce this great owl into the forests where it had become extinct. Thirty Swinhoe's pheasants were re-introduced in Taiwan in 1967 and released there as new breeding stock, and cheer pheasants were re-introduced in a forest reserve in India in 1971 and 1973. All were bred in the Pheasant Trust, which has its headquarters at Norfolk Wildlife Park.

The Pheasant Trust was started by a group of well-known ornithologists and aviculturalists in 1959 under the Honorary Directorship of Mr Wayre, with the object of trying to save the world's rarest pheasants from extinction, and it now maintains the largest collection of rare pheasants ever to have been assembled in the world. Several other release projects are planned for the future, which is an encouraging sign, for regrettably we note that the Trust is the only organization regularly carrying out work of this nature—a sobering fact to reflect upon when listening to the hot air frequently spouted by the zoo establishment on the subject of preservation!

Phillip Wayre's activities do not end with his work in the Wildlife Park and the Pheasant Trust, for, like the late Gavin Maxwell, Mr Wayre is a champion of the otter and on 22nd September 1971, he founded the Otter Trust with the object of promoting the conservation of otters whenever and wherever necessary for their survival. Sixty acres of land were purchased at Raydon in Suffolk in March 1972 as a site for the new project, but local opposition to the venture was so strong, basically on the grounds of the close proximity of the site to a local area of outstanding natural beauty, that Phillip Wayre withdrew his planning application in 1974 and began to search afresh for a suitable area. Fortunately a 23 acre site, bounded by a river, was eventually found and the Trust is now fully operational at Earsham, near Bungay in Suffolk.

Phillip Wayre's view of the zoo's duties seems abundantly clear, but should it require further elaboration, we quote from a letter he wrote to us in October 1976:

> So far as my own personal views are concerned I have always maintained that there is no moral justification whatever for keeping animals in captivity unless (a) every endeavour is made to breed them in order to reduce the strain on the remaining wild populations and (b) that they are kept in as near natural conditions as possible with every attempt being made to encourage young people to take an interest in wildlife and conservation through education. These are the aims of all three organizations which I run and as you know, two are already registered charities and the third will become so shortly. None is run with the object of financial gain.

Financial gain was not the motivation for the starting of the Wildfowl Trust either, which was founded by Sir Peter Scott in

1946 originally under the title of The Severn Wildfowl Trust. Slimbridge is the best known of their collections but equally well worth visiting are the waterfowl centres at Peakirk near Peterborough, Washington in Tyne and Wear and Martin Mear in Lancashire. The Trust is a non-profit-making charity whose object is to study the biology of wildfowl and some of its achievements are so well-known that even the layman seems conversant with them. Of these, the rescue of the Hawaiian goose is surely the most famous: so successful was the Slimbridge breeding programme that over 50 birds were sent to Hawaii to aid in rebuilding the wild population; but the Trust carries out equally important work with other species every day of the year. For instance, they have an extensive ringing programme with three ringing stations situated in Suffolk, Essex and Northamptonshire, and they constantly research into food and feeding behaviour, courtship, bird navigation, migration patterns and disease and its treatment.

At Slimbridge one will find the largest collection of wildfowl in the world, a resident population of 2,500 specimens and about an equal number of wild and semi-wild birds that fly into the grounds seasonally. In fact, of the 247 species and sub-species of wildfowl known to science, 170 can be seen here.

The Peakirk Wildfowl Refuge was established in 1957 mainly to act as a reserve colony and breeding centre for Slimbridge. Nevertheless it is an important collection containing over 100 species of waterfowl.

Washington was opened in May 1975, the aim being to specialize in North American species to emphasize the link with Washington DC.

None of these collections fall easily into the category of zoo, wildlife park or even bird collection, for they are totally unique wildfowl reserves or sanctuaries. Nor can the Wildfowl Trust be thought of as simply a Trust for captive propagation. Their research work is concerned with wildfowl in captivity and the wild, and they are active on both counts. The Trust in effect seems to have bridged the gap between the maintenance of wildlife in the wild and the maintenance of wildlife in captivity, a fact that is easily passed on to and assimilated by the visitor. Sir Peter Scott's lifelong association with wildlife is well known. He is a successful author, painter, broadcaster and conservationist, but the Wildfowl Trust must rank as one of his greatest achievements.

Of our selection of wildlife collections, we would be hard-pressed to choose a favourite: each has its own particular good and bad points, each its own remarkable achievements. But of all of them the one we think of as the supreme example of success against all odds is Jersey Zoo in the Channel Islands. This enthusiasm is partly engendered by the fact that one of the writers began his zoo career here and has watched its progress with constant interest ever since. But personal affection for the place is not needed to justify its existence, Jersey does that for itself.

The zoo was founded in 1959 by the naturalist and author of many best selling books on animals, Gerald Durrell, with a view to creating an entirely different kind of zoo, a zoo that would become a centre for research into breeding endangered species under captive conditions, a zoo that would become an Ark. Durrell was fully aware of the many problems of maintaining wild animals in captivity, of the psychotic disturbances and genetic degeneration so often found in zoo specimens and the difficulties in finding suitably balanced diets, but he felt certain that these problems were self-made. To his mind there was no such thing as a difficult species—misunderstanding of the animal by the zoo-man gave it this reputation. Technically it should be possible to cater for any species providing one understands its requirements. Even so, the grand dream was a gigantic undertaking, and anyone visiting the zoo during its first three or four years of operation might justly have considered Durrell's aspirations little more than a grand illusion.

The main problem, as always, was finance. In the early days the Jersey Zoo had a wide variety of popular species intended to appeal to the public ('A postage stamp collection' Durrell called it), in order to get enough money into the zoo to finance future developments. But although the attendance figures began to make it a viable operation it was soon clear that extra money had to be found. In 1963 Durrell established the Jersey Wildlife Preservation Trust, as a charity, to take over the collection and continue its objectives of scientific research and breeding of endangered species. Today nearly 6,000 people in 38 countries are Trust members and the importance of the Trust's work has become acknowledged throughout the world.

The park still has a long way to go before it assumes the shape envisaged in that original dream, but already it has reached the stage where the visitor, upon entering the gates, immediately realizes that he has found a very special sort of zoo. Perhaps the most striking feature of the park is the remarkable collection of some of the world's rarest animals: even the most fanatical of zoo-goers is certain to find some animal he has never seen before except in a book or photograph, many of which breed regularly. Top priority is given to controlled breeding of the rarer species. Forty-eight per cent of the mammals have been zoo-bred, and the park is noted for its breeding groups of lowland gorillas, orang-utans, lemurs and marmosets.

Most of the animal accommodation has been redesigned in recent years and, although some seems a little disappointing from the point of view of the standards one has come to expect of Jersey, the reptile-house appears to us to be the finest in the United Kingdom. Only half of this building is devoted to exhibition and the vivariums within this area are large by comparison with most zoos. The remainder of the building contains an office, breeding rooms, food preparation areas, incubators and so on; it is in fact a self-contained breeding/research unit.

There are numerous things the visitor should look for apart from the reptile-house and the apes, however. The only pair of Asian lions to be found in captivity in Europe are in this zoo. In the wild, this race is now confined to a single reserve in the Gir Forest in north-western India. The only bear to be found in South America, the rare spectacled bear, is to be found in this collection, and has been bred here. All the species of lemur are confined to Malagasy, where they are gravely threatened with extinction by extensive clearance of their forest habitat. At Jersey one can see the ring-tailed lemur, ruffed lemur, mouse lemur, fat-tailed dwarf lemur, mongoose lemur and the mayotte brown lemur. The last two species breed regularly in the park. There are two breeding groups of ursine black and white colobus monkeys.

The park has always endeavoured to stick to those species that it knows it can accommodate best, and these include many of the smaller species of mammals that are generally neglected by most zoos, as they tend to lack the glamour, or public pulling power, that these establishments seem to feel obliged to provide. Thus the park abounds with small mammals one is unlikely to see elsewhere—species like the spiny tenrec from Malagasy, which has bred regularly here for several years. Or the Jamaican hutia, a primitive rodent confined to the West Indies where they are severely threatened with extinction. They bred here for the first time in captivity. The crested rat from Ethiopia and Northern Kenya was also bred in captivity for the first time at Jersey.

A recent development is a new marmoset complex where four species of these delicate South American primates can be seen, the policy being always to keep only a few species but to keep them in viable breeding groups. Naturally several groups are breeding.

Nor have the bird species been neglected. Jersey has a fine record with these, too. Gold Coast touraco, grey touraco, Mexican jay and thick-billed parrot were all bred for the first time in captivity at the zoo. Perhaps the rarest bird here is the Congo pheasant, and Jersey is the only zoo apart from Antwerp to breed them.

Jersey can actually claim to have saved a species from extinction, for in 1969 they managed to obtain two pairs of white-eared pheasants, the first specimens to come out of China since 1936. At that time there were only 18 specimens known to be in the Western world and most of these were either interbred or past breeding age, so the Trust considered it vital that they established a breeding-pool of this species. The operation was plagued with disaster: the first cock bird died within 24 hours of arrival and a hen died because she got egg-bound. Even the surviving cock bird managed to injure its leg, yet incredibly it eventually mated and the Trust's thriving breeding colony was started. To date over 70 of these rare pheasants have been exported to collections (including Phillip Wayre's Pheasant Trust) in eleven countries, and all from this one breeding group. These pheasants are placed on deposit, remaining the property of the Trust, so as to ensure that in the event of a fatal infection or unforeseen disaster the captive pool could not be wiped out at a stroke.

We have not mentioned all the breeding successes of the Jersey Wildlife Preservation Trust, nor have we touched upon several of the rarities to be found in it, but it must be clear why we hold it in such high esteem, for all of this has been achieved in a brief 18-year period. And this is only the tip of the iceberg—a great deal is hidden beneath the surface. Jersey's scientific record system, for example, is rightly considered to be one of the best in the world—if not the best. In brief this consists of three separate cards—History, Medical and Behaviour—for every specimen kept in the collection, and on to these daily observations are recorded throughout the animal's life. Valuable information is already beginning to be gleaned from these on a variety of aspects of wildlife management and it is certain that similar, and equally comprehensive records should be kept by *every* organization caring for wild animals.

The Trust has also set up a nutritional research unit and has just completed building a new animal hospital, equipped with operating theatre, an office for the veterinary surgeon, a small pharmacy, a recuperation room, a post-mortem room, and a mess room with eating and sleeping facilities that will allow a keeper to remain with his animal day and night until it has fully recovered.

Gerald Durrell, having masterminded one of the greatest wildlife preservation centres in the world, has even more ambitious plans for the future, his total objective being to formulate the best methods of breeding and protecting wildlife in both the zoo and their natural habitat. These plans are detailed in the Jersey

Wildlife Preservation Trust's publicity leaflet, which invites those concerned with the future of wildlife to join the Trust. In this Durrell lists the Trust's aims and explains why they need financial support.

> To establish new colonies of gravely endangered species.
> To expand their ethological and biological programmes.
> To improve and enlarge the Veterinary Research Unit, with special reference to Artificial Insemination Techniques.
> To expand their Nutritional Research Programme.
> To conduct biological field studies of threatened species.
> To undertake rescue and re-introduction operations.
> To expand the publication of their scientific findings.
> To further the educational programme.
> To establish a 'University of Conservation and Controlled Breeding Techniques'.

There is no reason to doubt these declarations of intent. On its past record alone the Trust has established positive proof of its motives. And if this is not enough, Durrell makes it abundantly clear to Trust members exactly where priorities lie, for as members they receive a copy of the Annual Report and a regular newsletter.

Considering the progress to date on conservation and the determined attitude of the Park's Honorary Director and his staff, we look forward to the next ten years of its operation with pleasant anticipation.

This is the bulk of our selection, a mere handful of collections that are actively living up to the responsibilities of the modern wildlife collection, depressing statistics when set against the 148 we have found in operation at the end of 1976. Of these, 23 are bird collections, 10 safari parks, 16 dolphinariums and the remainder zoos in the general sense. We are not suggesting that all these are diabolical establishments, in fact some are extremely positive in their approach to wildlife management.

The Monkey Sanctuary at Looe in Cornwall, for example—without question the most unconventional zoo in the country—consists of only one species—a thriving breeding colony of woolly monkeys from South America. Mr Leonard Williams, who founded the Sanctuary in 1964, is hardly likely to thank us for using the word 'zoo' in connection with his collection, for he is a staunch critic of zoos, yet is on record as saying: 'Arguments against zoos in principle hardly merit serious consideration'.

This statement does not, as it might appear, contradict his many attacks on the British zoo, it merely endorses our own argument about them, for the zoo in principle is a fine institution that frequently falls short of the ideal. Leonard Williams' Monkey Sanctuary is an unqualified success simply because it represents the ultimate in specialization—one species is kept in large, grassed areas complete with beech trees, where they can behave in a completely natural way.

Woolly monkeys are not considered an easy species to breed in captivity and this is one of the few places in the world where regular breeding has been achieved, so a visit to the sanctuary provides the visitor with a true example of what keeping wild animals in captivity is about. Here they can learn about the species (an introductory talk on the life of the woolly monkey is given to all visitors), enjoy the sight of the monkeys living in an excellent environment and know that the animals are kept there for purposes other than amusement.

Specialization is, as far as we are concerned, the *only* way for the zoological collection to develop in the future and the reader will already have noticed that the collections we have selected all have a definite theme. Marwell, cats and ungulates: Howletts, gorillas, cats and ungulates; Norfolk, European birds and mammals; The Pheasant Trust and Otter Trust on those species; The Wildfowl Trust, water birds; Twycross, primates; The Monkey Sanctuary on woolly monkeys. Even Jersey's apparent general collection is refined to special themes; their policy is to keep those species most endangered but they have not attempted to obtain many of the larger animals, preferring to stick to the types less favoured for zoological exhibition. It is interesting to note that, being an island zoo, Jersey has begun to develop a collection of island species. A perfect idea, as island species rank as one of the most endangered groups of all: the potential for a study of island ecology with a view to eventual re-introduction presents an exciting prospect.

The conventional zoo with its excursions into every area of wildlife management cannot hope to compare with the specialist collection to any degree, and gigantic collections like London, despite its valuable behind-the-scenes research, merely seem to perpetuate the zoo image as found in so many children's books—camel rides, feeding the lions, a laugh at the funny monkeys. Progress will be really made when the mainstream zoo concept is shattered into dozens of pieces and the smaller, more specific collection begins to emerge into the foreground. The prospect is quite dramatic. Just think of the possible collections that might develop: a Herpetological centre perhaps, keeping only reptiles and amphibians, or an insect zoo concentrating on research into these fascinating creatures, a collection of rodents and insectivores, or a zoo that keeps exclusively marsupials.

We do not accept the standard zoo man's rejection of the idea of specialization, which is that not enough visitors would visit, say, a Herpetological centre, and that, deprived of financial support, it would flounder within the first twelve months. This might be true in the present climate of commercial exploitation where the public are constantly coaxed away from the better collections by the more tawdry; but the precedent has been set by collections like the Otter Trust where, apart from some waterfowl and a few mammals associated with the otter, all one can expect to see are otters. People are already going there and many are finding the experience a far more interesting type of outing than the usual zoo visit.

Naturally we are not suggesting that every collection should keep only one species, but that every collection should start to direct its attention towards specific objectives and become something more than an animated funfair, for the profit motive is undeniably the enemy of preservation.

We are forced to face the fact that the conservation lobby is beset with anti-zoo feeling, and some groups like the World Wildlife Fund appear to be actually hostile towards zoos. But when one considers the British zoo's record, the overt commercial exploitation and the meaningless cant mouthed by so many about conservation, can one really be surprised at this hostility? Conservation became fashionable, if that can be considered the right word, in the late 'fifties and early 'sixties, a period that coincided with the appearance of the profit-motivated zoo. The coincidence was most unfortunate, for the commercial groups profited handsomely from the increased attention that was being paid to wildlife and, as conservation became more and more a subject for discussion, their example gained the zoo a worsening reputation. If these zoos were closed down, it is indubitable that zoos and conservationists would quickly find many areas of common ground, and conservation and preservation by captive propagation would take an enormous leap forward. But how can this be done?

Legislation would not provide the total solution, if only because the British Government would not be able to differentiate between serious zoological collections and the commercial ones run for profit where animals are regarded as objects of entertainment. What it would achieve, though, would be beneficial. Certainly some of the less desirable collections would be forced to improve their standards or close down altogether, and that is something many of us would welcome.

Unfortunately, until the science-profit conflict is resolved, it seems that legislation on zoos will not be forthcoming in the United Kingdom. After all, in the first fight, Lord Janner's ill-fated Bill for Control of Zoological Gardens was soundly beaten by those interested in profit, not wildlife. This piece of legislation required that all zoological collections be registered with a proposed Zoological Council who would insist that certain conditions were adhered to. These were, namely, that the owner of the collection should make adequate provision for:

(a) The health and welfare of the animals in the collection.
(b) The provision of adequate veterinary care.
(c) That an adequate number of keepers are available at all times.
(d) The safety of the visiting public and of the staff.
(e) That the zoological garden is maintained in a proper state of repair.
(f) That proper records are kept of births and acquisitions and of deaths and disposals of animals.

One would not think that any of this was unreasonable, yet the bulk of zoos objected, they could not agree on the Council. Serious zoos did not want commercial zoo people elected to it; commercial zoos feared the election of academics. They could not agree on what each collection should pay in licence fees. A small collection clearly should not pay as much as a larger zoo, but should the latter have more say because it contributed more? Dozens of equally puerile objections were raised, constantly causing a diversion from the central issue. But the main source of these objections came from people who knew that, if the Bill became law, they would have to spend a sizeable sum of money to conform to its conditions; so they closed their ranks and scuttled a piece of legislation that was intended to protect the zoo animal from abuse.

In 1976 the Government did enact a Bill relating to wild animals, and it is a piece of legislation with unprecedented side-effects. This was the Dangerous Wild Animals Bill, which is intended to restrict the private ownership of wild animals considered to be potentially dangerous to man. Some of the species covered by this Act include all members of the cat family, apart from the domestic cat, all members of the dog family, again apart from the domestic, poisonous snakes, members of the crocodilians, primates, and so on. It was a law that was long overdue, the private ownership of species like lions having grown to ludicrous proportions; and today anyone wishing to own a specimen of any of the species listed has to obtain a licence. Strange as it must seem, this means that the private owner of a lion must conform to regulations relating to security and animal welfare that are non-applicable to the zoo-owner, animal-dealer or circus!

One has to conclude that legislation would be a step in the right direction, if not the complete answer. At least those organizations concerned with animal welfare would have a weapon to fight with—something they lack utterly at present. But in the final analysis the solution rests, as John Knowles suggested, in a change in public attitudes, in understanding what the operation is really all about.

The reader might be inclined to wonder what on earth can the man in the street do about the situation. He has the greatest power of all, for if he refused to patronize those places that simply exploit wildlife they would be forced to close. Identification of these establishments is not difficult either, as they can be divided into two groups: the obviously squalid menagerie and the slick, sophisticated operation that has as its objective straightforward financial gain. The former is easily recognized with its tatty cages and general air of neglect: if one happens upon one of these the thing to do is to complain to the management, the local council, the local MP and one of the animal welfare groups listed in the Appendix to this book. The more sophisticated operation usually gives itself away: read the guide-book carefully and ask questions. Anyone can say that the operation has serious objectives, the question is always, what exactly are they doing to realize them? If a charlatan is revealed—complain.

We shall doubtless be accused of rabble rousing, of irresponsibility, by exhorting the public to demand more of the zoo, but our efforts to bring the plight of many zoo animals to the attention of the public stem from a genuine concern for humane treatment of animals and a deep-rooted conviction that the zoo has much to contribute to wildlife preservation. We are not blind to the fact that the human population has more immediately pressing problems than the extinction of a number of species of animals, but a species gone is gone forever and it is clearly our duty to safeguard wildlife for future generations. The zoo can do much towards this goal—if it has a mind to.

For ourselves, we shall not desist from harassing any zoo that falls short of its duty, because we believe that no wild animal should be exploited simply to create an attraction or exhibit, and we would be greatly assisted in our task if every reader decided to apply pressure and make his local zoo a totally humane facility with positive aims—or close it. The only case of extinction we desire to see is that of the exploiter of wildlife.

# Appendix

Those who feel that they would like to help a programme of captive preservation can do so by joining one of the following organizations. Just send a stamped, addressed envelope for details.

The Wildfowl Trust,
Slimbridge,
Gloucester, GLT 7BT.

The Jersey Wildlife Preservation Trust,
Les Augres Manor,
Trinity,
Jersey,
Channel Islands.

Marwell Zoological Society,
Marwell Zoological Park,
Colden Common,
Winchester,
Hampshire.

The Hawk Trust,
Dept. A.,
P.O. Box No. 1,
Hungerford,
Berks, RM17 OPL.

The Pheasant Trust,
Great Witchingham,
Norfolk.

The Otter Trust,
Earsham,
Bungay,
Suffolk.

Twycross Zoo Association,
Twycross Zoo,
Atherstone,
Warwickshire.

Anyone who actually witnesses a case of neglect or cruelty should contact:

The Royal Society for the Prevention of Cruelty to Animals,
Manor House,
The Causeway,
Horsham,
Sussex, RH12 1HG.
or
The Captive Animals' Protection Society,
Flat 1,
17 Raphael Road,
Hove,
Sussex, BN3 5QP.

# Index

ERRATUM: FRONTISPIECE PHOTOGRAPH BY JOHN DOIDGE